PROTECTING ELECTRONIC EQUIPMENT FROM ELECTROSTATIC DISCHARGE

To Lorri, David, Jonathan, and Staci Richey

No. 1820
$16.95

PROTECTING ELECTRONIC EQUIPMENT FROM ELECTROSTATIC DISCHARGE

BY EDWARD A. LACY

TAB TAB BOOKS Inc.

BLUE RIDGE SUMMIT, PA 17214

Notice: Trimline® is a registered trademark of Bell Telephone.

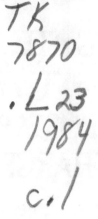

FIRST EDITION

FIRST PRINTING

Copyright © 1984 by TAB BOOKS Inc.

Printed in the United States of America

Library of Congress Cataloging in Publication Data

Lacy, Edward A., 1935-
    Protecting electronic equipment from electrostatic
discharge.

    Bibliography: p.
    Includes index.
    1. Electronic apparatus and appliances—Protection.
2. Electric discharges. 3. Electrostatics. I. Title.
TK7870.L23 1984   621.3815   84-8902
ISBN 0-8306-0820-6
ISBN 0-8306-1820-1 (pbk.)

Cover illustration courtesy of Larry Selman.

Contents

Acknowledgments   vii

Introduction   ix

1   Static Electricity, A Space-Age Gremlin   1

    1.1   Electrostatic Discharge
    1.2   Case Histories of ESD Damage
    1.3   Electrostatic Discharge—Your Problem
    1.4   A Short History of Static Electricity
    1.5   The ESD Environment
    1.6   Nature of Static Electricity
    1.7   Effect of ESD on Electronic Components
    1.8   Protection for Integrated Circuits

2   Principles of Electrostatics   20

    2.1   Electron Theory and Atomic Structure
    2.2   Electrostatic Attraction and Repulsion
    2.3   Triboelectric Charging
    2.4   Conductors and Insulators
    2.5   Charging by Induction
    2.6   Electric Fields
    2.7   Electrostatic Measurements: The Electroscope
    2.8   Electric Potential
    2.9   Equipotential Surfaces
    2.10   Surface Charge Density
    2.11   Capacitance
    2.12   Electrostatic Generators
    2.13   Static Charge Dissipation

3   Electrostatic Test Equipment   **53**

    3.1   Portable Electrostatic Meters
    3.2   Precision Electrostatic Voltmeters and
           Fieldmeters
    3.3   Electrostatic Monitors
    3.4   Surface Resistivity Probe
    3.5   Static Decay Meter
    3.6   Humidity Test Chamber

4   Protective Packaging   **69**

    4.1   Protective Bags
    4.2   DIP Tubes
    4.3   Protective Tote Boxes and Storage Bins
    4.4   Protective Foam
    4.5   Conductive Shunts

5   Protective Environment   **82**

    5.1   Humidity Control
    5.2   Work Surfaces
    5.3   ESD Protective Floors
    5.4   Air Ionizers
    5.5   Tools and Production Equipment
    5.6   Topical Antistats

6   Protected Worker   **100**

    6.1   Grounded Wrist Straps
    6.2   ESD Protective Clothing

7   The Complete ESD Control Program   **109**

    7.1   Setting Up and Monitoring an ESD Control
           Program
    7.2   Protected Areas
    7.3   General Guidelines for Handling ESDS Items
    7.4   Specific Area Handling Procedures
    7.5   Personnel Training and Certification
    7.6   Product Design
    7.7   Customer Responsibilities

Appendix   **123**

References   **134**

Glossary   **140**

Index   **161**

Acknowledgments

Many thanks to Burt Unger of Bell Laboratories for reviewing the entire manuscript and making invaluable suggestions and criticisms. Thanks also to Dan Anderson, Richmond Division of Dixico; John R. Curran, Stanley Vidmar; G. W. Kahler, Western Electric; John T. Conway, Monroe Electronics; Stanley Weitz, ElectroTech Systems, Inc.; L. P. O'Reilly, Intervinyls, Inc.; and Andy Thigpen for valuable assistance.

The Reliability Analysis Center, Rome Air Development Center, was instrumental in obtaining permission to reprint numerous drawings from authors of papers in the Electrical Overstress/Electrostatic Discharge Symposium Proceedings. Text references to these proceedings are abbreviated in the following fashion: EOS-1 = Electrical Overstress/Electrostatic Discharge Symposium Proceedings, 1979; EOS-2 = same symposium, 1980; EOS-3, 1981; and EOS-4, 1982.

Military Handbook DOD-HDBK 263, "Electrostatic Discharge Control Handbook for Protection of Electrical and Electronic Parts, Assemblies and Equipment," 2 May 1980, is such a basic reference that the text and drawings in it have been used liberally.

The following manufacturers and organizations graciously supplied information: ACL, Inc.; Anderson Effects, Inc.; Armand Manufacturing, Inc.; Bemis Company; Bengal, Inc.; BestWay/Araclean; Chapman Corporation; Charleswater Products, Inc.; Clean Room Products, Inc.; Colvin Packaging Products, Inc.; Conductive Containers, Inc.; Controlled Static;

Electronic Industries Association; Electro-Tech Systems, Inc.; Frontier Electronics, Inc., Gary Plastic Packaging Corporation; Glen-Mitch Tools, Inc.; Stephen Halperin V. Associates; Herbert Products, Inc.; Herbert Miller, Inc.; Jiffy Packaging; Julie Associates, Inc.; Lewisystems, Menasha Corporation; LNP Corporation; Meridan Molded Plastic; Merix Chemical Company; Misco, Inc.; Monroe Electronics, Inc.; NRD; Nu-Concept Computer Systems, Inc.; Protecta-Pack Systems; Republic Packaging Corporation; Richmond Division of Dixico, Inc.; Scientific Enterprises, Inc.; Sealed Air Corporation; Semtronics; Sentinel Foam Products; SIMCO Co., Inc.; Stanley Vidmar; Static Control Systems/3M; Static Handling, Inc.; Static, Inc.; Thielex Plastics Corporation; Trek; Vinyl Plastics, Inc.; Wescorp/DAL Industries, Inc.; Don White Consultants, Inc.

And thanks to my wife Rita for her support in ways too numerous to mention.

Introduction

Because of the development of semiconductor circuits, static electricity has become a major problem for the electronics industry. Even though you may not feel or see this gremlin, it is there, nevertheless, ready to zap sensitive devices. Whether you are a hobbyist or an engineer, it's important for you to know how this electrostatic discharge can affect your electronic equipment and what you can do for protection.

This book surveys the causes of electrostatic discharge and the equipment and procedures used to combat it. While some of the procedures and equipment apply only to large manufacturers, most of them can be scaled down to even the neighborhood television repair shop.

The purpose of this book is to convince all personnel involved in the manufacture, assembly, test, and repair of semiconductor circuits that (1) electrostatic discharge is a very real problem and (2) protection is available. In scope it is an introduction to electrostatic problems in the electronics industry and to their prevention. It avoids mathematical derivations but provides enough theory to give the necessary foundation for the practical tips that follow.

Chapter 1 defines the problem of electrostatic discharge and gives the reasons why this is a problem for everyone in the electronics industry, including the hobbyist. It shows that the protective networks supplied by IC manufacturers are not always adequate.

Chapter 2 gives the principles of electrostatics so that the reader can understand the problem of electrostatic discharge and how it can be controlled. Only that theory relevant to the field of electrostatic discharge is given.

Chapter 3 describes test equipment for detecting the presence of static electricity charge and measuring its magnitude and polarity. It includes the inexpensive meters suitable for a small shop and the elaborate precision meters needed for large manufacturers.

Chapter 4 shows how semiconductors should be protected with special shipping bags, tubes, and foams. It also describes temporary containers to be used during manufacture and assembly.

Chapter 5 gives ways for making the environment safer for semiconductors. These methods are in addition to protective packaging, but not a substitute for them.

Chapter 6 shows how the worker himself should be dressed as still another way of preventing electrostatic discharge.

Chapter 7 summarizes the preceding chapters in order to show the complete program needed to control electrostatic discharge.

The appendix provides definitions and addresses that the newcomer to electrostatic discharge control needs but may have trouble locating.

In this book the reader will learn:

- How to choose electrostatic test equipment
- Techniques for handling semiconductors
- How to select the proper air ionizer
- How to protect semiconductors during shipment and storage
- How to use tools around semiconductors
- What to consider in selecting wrist straps
- How to use humidity control in electrostatic discharge control
- How to build a static-free work station
- How to use electrostatic discharge control techniques in the field.

While these procedures can be time-consuming and protective equipment can cost a few dollars, they are well worth it in order to use the exciting new devices that are being developed.

1

Static Electricity, A Space-age Gremlin

I N RECENT YEARS IT HAS BECOME OBVIOUS THAT THERE HAS BEEN A dramatic increase in the use of electronic equipment in everyday life. From word processors in the office, to robotics in the factory, to personal computers in the home, sophisticated electronic devices have become an integral part of most of our lives. Along with the electronic cottage, we now have the electronic automobile and the electronic office and factory. This has come about almost entirely due to the advances that have been made in semiconductor technology, from the invention of the transistor, to the integrated circuit, and now to the microprocessor.

Since the invention of the integrated circuit, it seems that each year more and more transistors have been crowded into individual ICs so that now a complete system on one silicon chip is commonplace. Because of the development of *very large scale integrated circuits,* it's now possible to purchase electronic equipment that only a few years ago would have been prohibitively expensive.

Even though a typical IC may have many thousands of transistors, IC manufacturers keep trying to crowd more and more transistors on a single chip by microminiaturization. At the same time they are trying to increase the operating speed of ICs and to reduce power consumption. To accomplish this, they are making gate oxides (dielectrics) thinner, junctions more shallow, and interconnecting lines finer. For example, PMOS and NMOS ICs have a channel length of approximately 10 microns (1 micron = 1 millionth of a meter) and an oxide thickness of 1100 to 1500 angstroms (1 angstrom = 0.0001 micron).[1] The trend in size reduction is shown rather dramatically in Fig. 1-1. Devices under development have a quarter-micron channel length and dielectrics that are only 150 angstroms thick!

But each time manufacturers shrink chip geometries, the ICs become that much more sensitive or vulnerable to destruction or damage from that common, age-old phenomena of static electricity or electrostatic discharge. That stinging shock you feel as you touch the door handle as you slide across the seat of your automobile on a winter day is the same phenomenon that has zapped ICs around the world.

1.1 ELECTROSTATIC DISCHARGE

While electrostatic discharge (ESD) has been a problem with MOSFETs from the time of their invention, the problem is becoming worse with each increase in the complexity of a chip. The thinner the oxide, the less the breakdown voltage. Because of the severity of the problem, reliability engineers describe ESD as our *most significant* electronic component problem, in effect, an electrical saboteur.

Now it has been found, much to the surprise and consternation of technicians and engineers around the world, that ESD is a problem for many devices other than MOSFETs. Surface acoustic wave (SAW) devices, operational amplifiers, JFETs, SCRs, microwave semiconductors, thin film resistors, Schottky diodes, some ECL and TTL circuits, resistor chips, and piezoelectric crystals are also ESD sensitive (ESDS) devices. In some cases, even relays, connectors, and printed circuit boards are vulnerable.

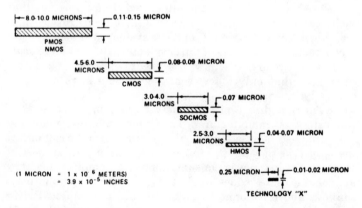

Fig. 1-1. Typical gate oxide cross sections—reduction to improve circuit density and response time constraints (courtesy of Donald E. Frank[2]).

2

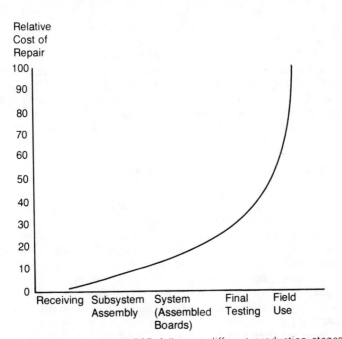

Fig. 1-2. Estimated cost of ESD failure at different production stages (courtesy of George R. Berbeco[10]).

Once installed in a subassembly or equipment, the ESDS device *may be* safer than it was before installation. *But* it may be even more sensitive if PC board circuits act as antennas to gather even greater voltages to zap the ESDS devices.[3] Thus, it's best to assume all electronic components and assemblies are sensitive to ESD damage.[4]

It's hard to say precisely how much ESD is costing us, but there's not much doubt that it's tremendous. Estimates range as high as $500 million a year[5] which seems unbelievable until you consider (1) how many millions of ICs are being produced each year and (2) the cost of ESD damaged devices includes the cost to find and replace them. See Fig. 1-2.

For all component failures it has been estimated that between 5 percent to 25 percent are ESD caused;[6] for dead-on-arrival components, 50 percent.[4] Greater than 50 percent of early operating failures are caused by ESD.[7] One manufacturer estimated that 60 percent of its field service calls were related to ESD damage.[8] One Western Electric plant reported a 2300 percent return on its ESD control investment.[9]

ESD is a problem for everyone in the electronics industry: from the semiconductor manufacturer, to the manufacturer who assemblies it into equipment, to the technician who repairs or

3

operates the equipment in the field. It affects the home electronics buff as well.[5]

In fact, everyone involved in processing, assembly, inspection, handling, packaging, shipping, storage, stowage, testing, installation, and maintenance of susceptible parts throughout the equipment life cycle, both at the manufacturer's and the user's facility needs to be aware of the damage he or she can do to ESDS parts.[11] Let's look at some case histories of damage from electrostatic discharge.

1.2 CASE HISTORIES OF ESD DAMAGE

A video game manufacturer had above normal returns during winter months because of a constant firing mode problem that was attributed to static electricity.[12]

At a large instrument manufacturer, poor handling procedures in the components area resulted in ESD damage to over 30 percent of the boards of one PCB assembly department.[13]

It was found at a bowling alley that bowlers moving across waxed wooden floors usually on rubber soles would zap automatic score-keeping equipment. When a conductive transparent coating was placed on the top surface of the score card and a grounding circuit to match was placed on the equipment, the problem was eliminated.[14]

At Sencore Corporation an unexplainable increase in solid-state device parts failures on production lines and an increase in stockroom parts failures were traced to ESD.[15]

During field tests of Bell Telephone's Trimline® telephone several instruments had failures of a thin-film circuit that provided the tones for dialing. It was found that when people walked across a carpet to pick up the phone, they would collect an electrostatic charge on their bodies. When they picked up the handset the charge was released into it and the thin-film circuit was destroyed. Once the problem was identified, Bell engineers quickly resolved it by adding a metal shield behind the faceplate of the instrument, thereby protecting the sensitive circuitry.[16]

At General Telephone Company, one Wisconsin field area was experiencing a higher than average failure rate for a particular piece of equipment. When it was found that static control procedures were not being followed, necessary changes were made and the failure rate fell back to normal. In another case at General Telephone a repair shop was having high failure with a CMOS electronic tone generator used in push-button instruments. Once again, when static controls were put into effect, the failure rate dropped dramatically.[17]

However interesting these case histories may be, there is a tendency to minimize them, to believe that electrostatic discharge is a problem for others, not the reader. To understand this rationalization let's look at the next section.

1.3 ELECTROSTATIC DISCHARGE IS YOUR PROBLEM

While some facilities, some shops, some locations may have more severe problems with electrostatic discharge than others, it is still *your problem* if you are engaged in any form of electronic equipment manufacture, test, or repair. You may be tempted to deny its existence or its magnitude because:

1. You can't see or feel any static discharge in your work location.
2. You believe ESD is a problem only with CMOS.
3. You believe that protective circuits provided by IC manufacturers are sufficient.
4. You believe once a part is installed on a board or module there is no longer a problem with ESD.
5. Failures are attributed to other causes.
6. You don't want to pay the price in dollars and inconvenience for ESD control.

In many ways, static electricity is like a space-age gremlin. That is, in many situations it can cause damage without you being able to see its spark, feel its sting, or hear its crackle. But you are not likely to feel ESD unless it is in the range of 2000 to 4000 volts, depending on the person, or higher. Yet many, if not most, semiconductor devices are sensitive to static well below this threshold of perception, that is, they will fail at voltages less than 2000 volts if zapped by ESD. By the time you hear an audible crack or see a spark, the electrostatic discharge is much higher.

With the instruments listed in Chapter 3 you can make your own measurements to determine the magnitude of your electrostatic problem. If a gross indicator will suffice you can use a common NE-2 neon bulb in a technique popularized by Dan Anderson to detect static charges. In this procedure the leads of the bulb are spread out in the form of a dipole. The bulb is then placed near an item suspected to have a static charge. It's best to do this test in a darkened room.

Actually, no tests are necessary to show the existence of harmful static electricity in your work area; just about all work areas around the country will show such voltages at some time or another during the year.

For a precise survey of your static electricity problem you can call in ESD control experts who will measure static voltages at pertinent points. In addition to making measurements, they will also make recommendations for ESD control, all for a fee that may be as much as $1000 a day.

Even when static charges are shown in an area, some will blame device failures on power line transients or improper hookup, test, or assembly.[3] The only sure way to prove ESD damage is to send a failed device to a specially equipped laboratory for failure analysis. In this procedure the device is disassembled and placed under a microscope for examination. As the conventional microscope has a limited optical resolution, it's necessary to use a scanning electron microscope. With such a microscope you can see tiny pinholes that ESD has punched through the dielectric and also leakage paths caused by ESD. Unfortunately, such microscopes are quite expensive and therefore not at all common. The complete failure analysis procedure can be difficult and lengthy, and therefore expensive.

Even with sophisticated equipment, it's very hard to be 100 percent sure that ESD is the cause of component damage. A very practical test in the factory is to run A/B comparisons. Lot A is the normal lot; lot B run at the same time is handled with all the static precautions. If the lots are large enough and the tests sufficiently discriminating, the difference in yields will be apparent. Also, comparing winter and summer yields will generally show a cyclic pattern associated with ESD.

If you are in the factory and having no problems, you may not know about problems in the field. If field technicians and customer service engineers do not have a static awareness, they may not be communicating ESD problems to the factory. This may happen if field technicians do not return failed parts to the factory for analysis.

Make no mistake about it, if you have a problem with ESD, the solution can cost you money as well as convenience. You, or those who work for you, may have to change your work habits. For instance, you may have to start wearing grounded wrist straps. You may have to change your environment—some areas will have to be restricted access. Static control materials and equipment will have to be installed. Workers will have to be sold on the idea, and resold. Static awareness, which we will discuss in a later chapter, will have to be routine. Some may feel, erroneously in most cases, that the cure for ESD may cost more than the damage from ESD.

Once you decide to proceed with ESD control, you can use the later chapters in this book as a guide for your ESD control program. But before we consider the sources of ESD, let's digress and review the history of ESD control.

1.4 A SHORT HISTORY OF STATIC ELECTRICITY

Except for lightning, the ultimate in electrostatic discharge, ancient man probably had no problem with ESD unless it may have triggered gunpowder explosions. But by the 1930s, ESD was being blamed for explosions in mines and in hospital operating rooms. With the widespread use of plastics after World War II, ESD became even more of a problem. New gases were introduced in hospital operating rooms to replace the highly combustible ones that had been used. Explosions continued, however, in grain elevators and in manufacturing operations where flammable vapors were produced through the use of flammable coatings and solvents. ESD caused inadvertent firing of electroexplosive devices as the trigger devices could not distinguish between an ESD voltage and a conventional triggering voltage. These problems continue to this day, and more and more problems are still being discovered with static electricity.

In manufacturing processes static interferes even when fires or explosions do not result. Powders resist mixture, textile fibers cling, repel, or break when not desired, and paper and plastic film cling at the wrong time. All of this occurs because of high speed production machines, the proliferation of synthetic materials, and the use of very thin paper and film.

Because static electricity attracts dust, it creates problems of contamination in the manufacture of printed circuit boards and optical devices. As it attracts dust, it gives a poor appearance to customer packaging.

Despite all these problems, there is a good side to static electricity. It has been put to very effective use in copying machines and some painting processes. Static electricity is used in the graphic-arts industry to make direct and indirect screens without using a vacuum frame, in metal working to hold down objects to electrostatic chucks, and in the electrical industry to test the integrity of cable insulation.[18]

By 1960 scientists had pointed out that ESD could harm semiconductors. An article in the January 12, 1962 issue of Electronics Magazine put it bluntly: "Static electricity can kill transistors." In 1966 the Richmond Corporation developed antistatic pink polyethylene because of a disastrous missile ignition at Cape Canaveral, caused by ESD. This material was later to be used extensively for ESD control in the electronics industry, although this application was not obvious at the time.

In 1969 a manufacturer of precision thin-film resistors became aware of ESD after hundreds of their resistors were degraded by ESD in shipment.

In the early 1970s, failure analysis techniques too often

depended on conventional microscopes and thus much ESD damage to semiconductors went unnoticed. As the use of scanning electron microscopes spread, ESD damage became more noticeable.[19]

By 1975 the Jet Propulsion Laboratory and Hughes Aircraft, to name just two organizations, had started preparing procedures on how to protect microcircuits from ESD. Within a year or two, seminars on electrostatic discharge were being conducted by the Reliability Analysis Center.

In 1979 the first annual Electrical Overstress/Electrostatic Discharge Symposium was held. It was sponsored by the ITT Research Institute.

In 1980 the Department of Defense published two important and very significant documents: DOD Handbook 263 and DOD Standard 1686. These documents were prepared by the Naval Sea Systems Command to provide guidance in developing, implementing, and monitoring elements of an ESD control program.

The EOS/ESD Association was formed in 1982 with its objective to "Strive for advancement of theory and practice of electrical overstress avoidance . . . The field of interest of the Association shall be the design hardening and prevention aspects of electrical overstress. This especially includes phenomena of electostatic discharge and its control as applicable in design, manufacturing, and end use."

Present members of the EOS/ESD Association represent most of the major electronics companies in the United States.

1.5 THE ESD ENVIRONMENT

ESD damage is much more likely to occur in dry climates than in wet ones. For example, Phoenix, Arizona with a low of 12 percent relative humidity has a bigger problem than cities such as Los Angeles where the outside relative humidity rarely drops below 45 percent. Notice that these are *outside* readings.

When heating systems are turned on in the winter months, the *inside* humidity may drop to disasterously low levels even in Los Angeles. Inside relative humidity readings of 25 percent are not uncommon during such times. During such conditions losses from ESD noticeably increase. '

As we noted earlier, ESD is a continuous threat to electronic components, from the time a device is manufactured, through its operation and maintenance of the equipment in the field. That is, ESD is not simply a problem in manufacturing plants; it's a problem in the neighborhood repair shop and in the shop or office where a customer engineer is making a repair or adjustment. Probably no one in this cycle is immune to ESD problems.

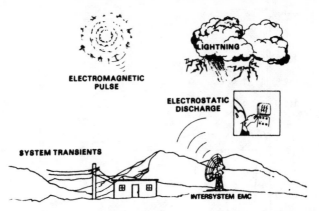

Fig. 1-3. Electrical transients are produced by many sources (courtesy of D. G. Pierce[20]).

Shipping clerks, assembly line workers, technicians—all may contribute to the problem. As a result a device may be zapped numerous times before shipment.

Electronic components and semiconductors in particular must exist and survive in a hostile environment where transient electrical overstress can routinely come from many directions. As shown in Fig. 1-3, these transients include the rare (but disastrous) electromagnetic pulse from a nuclear explosion, intersystem electromagnetic interference, lightning, system transients, and electrostatic discharge.

While some of these transients may last just a few nanoseconds, as shown in Fig. 1-4, they can still damage or destroy fragile semiconductor junctions. As ESD is the biggest problem, this text is devoted entirely to ESD.

1.6 NATURE OF STATIC ELECTRICITY

Static electricity, as you recall, is simply an excess or deficiency of electrons on a surface. It most often occurs when two materials come together, either through rubbing or simple contact, and then are separated, as when components slide around in plastic shipping containers. (But it can also happen with gases and liquids.) After the materials are separated, one will have a positive electrostatic charge, the other negative. The polarity and size of the charge will be determined by the speed and duration of motion and by the types of materials involved. Plastics and synthetic fibers are the worst offenders.

Static electricity is also generated by movement of the human body as when you walk across a carpet or slide off an upholstered chair, or rub your arm across the top of a workbench.

9

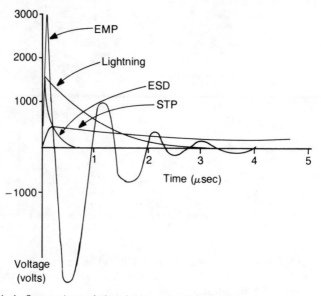

Fig. 1-4. Comparison of electrical transients for various threats (courtesy of D. G. Pierce[20]).

The static charge level on a persons's body has been measured as high as 35,000 volts when the relative humidity was very low. But even at 55 percent relative humidity charges as high as 7,000 volts have been measured on the human body.

If a charged person or object comes into contact with an electronic component or device, the subsequent electrostatic discharge may destroy the component. In some cases, direct contact may not be necessary for damage to occur. An induced electrostatic field may also cause havoc. We will discuss generation of electrostatic charges more thoroughly in Chapter 2.

1.7 EFFECT OF ESD ON ELECTRONIC COMPONENTS

When a static-charged person or object gets near an ESD sensitive component, trouble may result either from direct contact between the person or object and the component, from indirect contact as when a person touches a printed circuit board that has an ESDS component on it, or from the electric field that is induced on the component.

Just how much the part will be damaged or degraded depends on its sensitivity or susceptibility to ESD. This sensitivity is typically expressed as a voltage level that will cause catastrophic damage, in effect an electrostatic voltage threshold. Table 1-1 shows three classes of ESD sensitivity: class 1: items susceptible to damage from ESD voltage levels of 1,000

Table 1-1. Classification of ESD Sensitive Parts (DOD-HDBK 263).

Class 1: Sensitivity Range 0 to ≤1000 Volts

Metal Oxide Semiconductor (MOS) devices including C, D, N, P, V and other MOS technology without protective circuitry, or protective circuitry having Class 1 sensitivity.
Surface Acoustic Wave (SAW) devices
Operational Amplifiers (OP AMP) with unprotected MOS capacitors
Junction Field Effect Transistors (JFETs) (Ref.: Similarity to MIL-STD-701: Junction field effect, transistors and junction field effect transistors, dual unitized)
Silicon Controlled Rectifiers (SCRs) with Io<0.175 amperes at 100° Celsius (°C) ambient temperature (Ref.: Similarity to MIL-STD-701: Thyristors (silicon controlled rectifiers)
Precision Voltage Regulator Microcircuits: Line or Load Voltage Regulation <0.5 percent
Microwave and Ultra-High Frequency Semiconductors and Microcircuits: Frequency >1 gigahertz
Thin Film Resistors (Type RN) with tolerance of ≤0.1 percent; power >0.05 watt
Thin Film Resistors (Type RN) with tolerance of >0.1 percent; power ≤0.05 watt
Large Scale Integrated (LSI) Microcircuits including microprocessors and memories without protective circuitry, or protective circuitry having Class 1 sensitivity (Note: LSI devices usually have two to three layers of circuitry with metallization crossovers and small geometry active elements)
Hybrids utilizing Class 1 parts

Class 2: Sensitivity Range >1000 to ≤4000 Volts

MOS devices or devices containing MOS constituents including C, D, N, P, V, or other MOS technology with protective circuitry having Class 2 sensitivity
Schottky diodes (Ref.: Similarity to MIL-STD-701: Silicon switching diodes (listed in order of increasing trr))
Precision Resistor Networks (Type RZ)
High Speed Emitter Coupled Logic (ECL) Microcircuits with propagation delay ≤1 nanosecond
Transistor-Transistor Logic (TTL) Microcircuits (Schottky, low power, high speed, and standard)
Operational Amplifiers (OP AMP) with MOS capacitors with protective circuitry having Class 2 sensitivity
LSI with input protection having Class 2 sensitivity
Hybrids utilizing Class 2 parts

Class 3: Sensitivity Range >4000 to ≤15,000 Volts

Lower Power Chopper Resistors (Ref.: Similarity to MIL-STD-701: Silicon Low Power Chopper Transistors)
Resistor Chips
Small Signal Diodes with power ≤1 watt excluding Zeners (Ref.: Similarity to MIL-STD-701: Silicon Switching Diodes (listed in order of increasing trr))
General Purpose Silicon Rectifier Diodes and Fast Recovery Diodes (Ref.: Similarity to MIL-STD-701: Silicon Axial Lead Power Rectifiers, Silicon Power Diodes (listed in order of maximum, dc output current), Fast Recovery Diodes (listed in order of trr))
Low Power Silicon Transistors with power ≤5 watts at 25°C (Ref.: Similarity to MIL-STD-701: Silicon Switching Diodes (listed in order of increasing trr), Thyristors (bi-directional triodes), Silicon PNP Low-Power Transistors (Pc ≤5 watts@TA = 25°C), Silicon RF Transistors)
All other Microcircuits not included in Class 1 or Class 2
Piezoelectric Crystals
Hybrids utilizing Class 3 parts

Device Type	Range of ESD Susceptibility (Volts)
VMOS	30 — 1,800
MOSFET	100 — 200
GaAsFET	100 — 300
EPROM	100
JFET	140 — 7,000
SAW	150 — 500
Op-amp	190 — 2,500
CMOS	250 — 3,000
Schottky Diodes	300 — 2,500
Film Resistors (Thick, Thin)	300 — 3,000
Bipolar Transistors	380 — 7,000
ECL	500*— 1,500
SCR	680 — 1,000
Schottky TTL	1,000 — 2,500

* PC board level

volts or less; class 2: items sensitive to voltage levels of 1,000 to 4,000 volts; and class 3: parts sensitive to voltages of 4,000 to 15,000 volts. These voltages are based on a standard test circuit, which we will discuss in Chapter 2. If different test circuits are used, different values will be obtained.

Table 1-2 expresses ESD sensitivity in a little different manner, showing the parts sensitivity in a relative fashion. It is obvious from this table that some parts have a greater need for protection from ESD.

Damage to these parts is caused primarily by junction burnout, dielectric breakdown, and metallization melt. In *junction burnout* (also known as thermal secondary breakdown or avalanche degradation), a hot spot is formed in a semiconductor junction by a constant voltage pulse. If the pulse is long enough, the silicon in the hot spot will melt and short the junction, as shown in Fig. 1-5.[20]

Dielectric breakdown (oxide punchthrough), shown in Fig. 1-5, is a common type of failure in MOS parts. It occurs when the voltage applied across the dielectric exceeds the breakdown voltage characteristic of the material. For MOS integrated circuits, this dielectric typically has a breakdown voltage of approximately 8×10^6 volts per centimeter, which converts to 80 volts or less for a 1000 Å dielectric.[22]

When the dielectric is punctured, current flows, destroying the oxide and forming a short. In some cases, however, the ESD pulse may not have sufficient energy to short the oxide; instead the device may recover or heal from such a puncture and continue to operate. It may fail later, unfortunately.

Metallization melt or burnout occurs when ESD transients increase the temperature of the part sufficiently to melt metallization strips or to fuse bond wires. As shown in Fig. 1-5, an open circuit then results.

Other causes of ESD failure are gaseous arc discharge (in parts with closely spaced electrodes the arc discharge causes vaporization and metal movement which degrades performance), surface breakdown (in perpendicular junctions this is a localized avalanche multiplication process which results in a high leakage path around the junction, putting the junction out of use), and bulk breakdown (high local temperatures within the junction area cause changes in junction parameters caused by formation of a resistance path across the junction).

But how can we prove that one of these failures has occurred? One common technique is to observe the electrical characteristics (current-voltage) of a suspected microcircuit on a curve tracer. Figure 1-6 shows these characteristics for a shorted junction, a degraded junction, and a normal junction.

Still further proof may be obtained by examining the device with a scanning electron microscope. With this instrument, it's possible to see the miniature craters that have been blown in the device, as shown in Figs. 1-7, 1-8, and 1-9. Note that a conventional microscope may not be powerful enough to reveal such faults.

The failures that result from ESD can be classified as either immediate or delayed, as temporary or permanent, as direct or

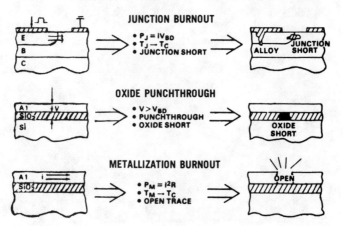

Fig. 1-5. Electrical overstress induced failure mechanisms (courtesy of D. G. Pierce[20]).

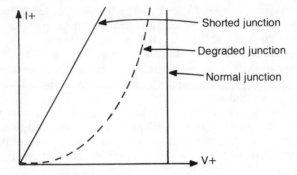

Fig. 1-6. Junction current-voltage trace (courtesy of W. Y. McFarland[9]).

indirect, or as upset or catastrophic (hard) failures. Some of these types, it should be noted, overlap.

In an *immediate* failure, the results can be observed within seconds of the time of the electrostatic discharge. In a *delayed* failure, commonly called a *latent* failure, a device shows no

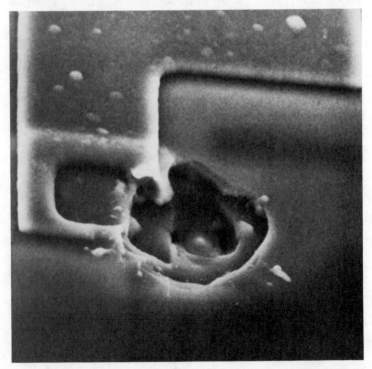

Fig. 1-7. Crater formed on surface of an integrated circuit by a 2000-volt charge (© 1982 Bell Laboratories Record).

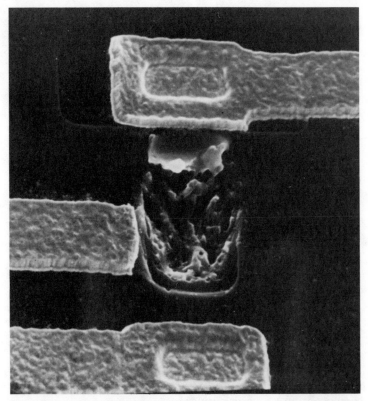

Fig. 1-8. Melting and cratering of a bipolar-type semiconductor (© 1982 Bell Laboratories Record).

immediate symptoms of failure, of being out of specification, after being exposed to ESD, yet it later fails sooner than identical parts that were not exposed to ESD. Such parts have been called "walking wounded"[1] or "pregnant with failure."[23] Whatever it is called, it is a serious matter for reliability engineers.

In the *upset* (also called intermittent or temporary) failure there is no apparent hardware damage but there is a loss of information or temporary distortion of the equipment's functions. After the ESD exposure the equipment generally resumes operation automatically although in the case of some digital equipment it may be necessary to resequence the equipment. In contrast, a *catastrophic* (or permanent) failure is an irreversible event, resulting in a failure that must be repaired before the equipment can be resequenced. A catastrophic failure can be either immediate or latent. Note that upset failures occur only when the equipment is operating while catastrophic failures can occur at any time, whether the equipment is operating or not.

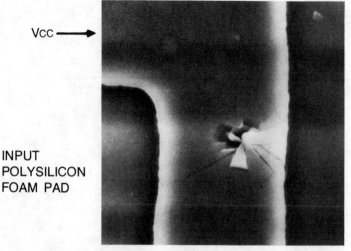

Vcc ⟶

INPUT
POLYSILICON
FOAM PAD

Fig. 1-9. Breakdown of the insulating layer in a metal oxide semiconductor (© 1982 Bell Laboratories Record).

In *direct* failure, known as a system hard error, part of a device is destroyed. In contrast, *indirect* failure, a system soft error, does not cause permanent damage.[24]

Still another method of classifying failures is either *degradation* failure or *catastrophic* failure. In the degradation failure a parameter of the component has shifted outside its specified range whereas the catastrophic failure is one in which the component no longer performs.[25]

Regardless of the type of failure, the obvious question is, Why can't IC manufacturers build protection circuits into their devices and eliminate the entire ESD problem?

1.8 PROTECTION FOR INTEGRATED CIRCUITS

For several years attempts have been made, and some have been partially successful, to protect ICs with off-chip transient suppressors and with on-chip protection networks. Let's consider some of the problems with these solutions.

1.8.1 Transient Suppressors

Suppressors to limit transients before they reach an IC are available in two common types: metal oxide varistor and silicon pn junction (such as General Semiconductor's TransZorb). The current through these devices is extremely small at low voltages. At voltages above a set clamping voltage, the current increases rapidly; the resultant voltage drop through the source impedance reduces the voltage peak.[26]

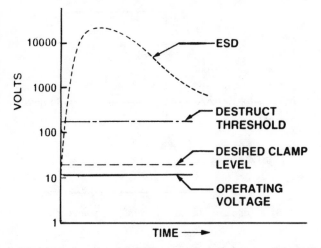

Fig. 1-10. Ideal clamping for ESD protection (courtesy of O. Melville Clark[31]).

However, such suppressors have very large parasitic capacitances which are unacceptable for some systems. Suppressors are being developed which will have a lower parasitic capacitance. However, the newer suppressors may still be impractical because of the extra parts, extra wiring, cost, and increased board space that would be required to protect each gate input on a printed circuit board.[26]

1.8.2 Protection Networks

For several years some sort of protection against ESD has been built into ICs, particularly MOS. But these devices or

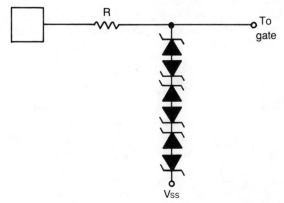

Fig. 1-11. Zener diode gate protection network (DOD-HDBK 263).

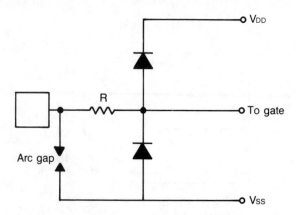

Fig. 1-12. Spark gap and diodes gate protection network (DOD-HDBK 263).

networks have not been sufficient because of numerous problems.

In general, most of the protection networks are built into the input leads of the ICs, with no protection for the output. Yet numerous ICs have been destroyed or degraded by electrostatic discharge through the output leads, which have no protection.[9] Fortunately, at least one manufacturer has both input and output protection networks for their ICs.[27] Still another problem with some protection networks is that the protective networks fail because of ESD and thereby take the IC out of operation.[28],[29]

In the ideal protection circuit, high voltages see a low impedance path around a gate but low voltages are not affected by this path. It has a fast response between high and low impedance conditions, yet it does not degrade the speed of the device it protects.[30] It provides the clamping shown in Fig. 1-10.

One protection circuit that has been used is a combination of Zener diodes, as shown in Fig. 1-11. As Zeners require greater than 5 nanoseconds to switch, they may not be fast enough to

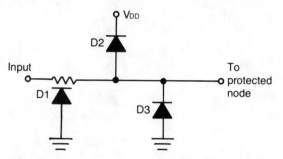

Fig. 1-13. Resistor-diode protection circuit (courtesy of Jack E. Keller[22]).

18

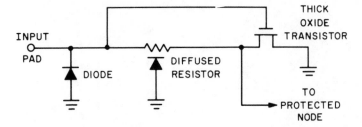

Fig. 1-14. Thick oxide enhanced punch through transistor (courtesy of Jack E. Keller[22]).

protect an MOS gate.[11] Still another gate protection network is the spark gap used in conjunction with diodes, as shown in Fig. 1-12.

Resistor-diode combinations have been a popular type of protection circuit, as shown in Fig. 1-13. Notice in this circuit that D1 is a diffused resistor, which in effect becomes a diode as well as a resistor. As a resistor, of course, it limits current flow. As a diode it protects the gate because its reverse breakdown voltage is less than the breakdown voltage of the gate.[22]

According to one researcher,[22] the most effective protection network for normal NMOS or PMOS is the thick oxide enhanced punch-through transistor shown in Fig. 1-14. It incorporates several of the best features of various protection networks.

Protection networks can adversely affect the performance of the device they protect, can take up too much chip space, and provide too little protection (that is, only up to 2,000 volts (2 kV) in some cases where several thousand volts protection may be needed).

Because of the push to put more and more on a chip, protection networks are being pushed to be made smaller. With careful design, manufacturers should be able to raise the threshold voltage to 2 kilovolts and even higher. Some manufacturers have achieved threshold voltages of 6 to 8 kilovolts and even higher. But even a level of 2 kilovolts is not likely to be universally provided until users insist on this level of protection.[23]

2

Principles of Electrostatics

To UNDERSTAND THE DEVICES AND TECHNIQUES FOR CONTROLLING electrostatic discharge, we need to master the basic principles of electrostatics—the science of electric charges at rest. Though some of the material presented in this chapter may seem elementary to those who are involved in electronics, we will introduce some concepts not normally included in such discussions.

Long before electrostatic discharge became a problem for the electronics industry, the study of stationary electric charges—static electricity—was a fundamental part of basic courses in physics. Before any student began a study of electricity and magnetism, he or she was expected to master the basics of electrostatics.

For many students, static electricity was merely a mildly interesting phenomenon, which became apparent on cold winter days. True, it helped to explain the structure of the atom and the more useful, more exciting phenomenon of electric current— charges in motion. And it was recognized as the cause of terrific explosions in mines, munitions factories, flour mills, and hospital operating rooms. But, still, it did not receive much attention until the development of integrated circuits, particularly MOSFETs. Now electrostatics has become a vital concern of electronics engineers.

It all began more than 2600 years ago when some Greeks observed that a piece of amber, a hard yellowish or brownish translucent fossil resin, rubbed with fur or flannel would attract light objects such as paper. For 2000 years or so, this phenomenon was ignored until the English physicist William Gilbert in 1600 studied the matter more thoroughly. He showed that the same result could be obtained with substances other than amber. To describe this condition, he coined the word *electrified,* from the Greek word for amber, *elektron.* But it was not until the 18th century that much was known about electrified bodies—bodies with electric charges.

Electrostatics is generally concerned with stationary electric charges, ones that are at rest or that occasionally move around at random. When these charges move, they form electric currents similar to other electric currents except that these currents are short-lived transients, seemingly instantaneous. Because of this short life, such currents cannot be harnessed and put to use for operating our myriad of electric devices and gadgets. They are not continuous currents.

2.1 ELECTRON THEORY AND ATOMIC STRUCTURE

To understand the nature of electric charges, we must examine the atomic structure of matter. As you recall from high school science classes, all matter is composed of very small particles called *atoms.* An atom is the smallest division of an element you can have and still be able to identify the element.

While tiny beyond imagination, the atom can be subdivided into three subatomic building blocks: protons, electrons, and neutrons.) While it may be an oversimplification, it is nevertheless safe to assume that these particles are arranged in the form of a solar system where the electrons correspond to the planets and the protons and neutrons correspond to the sun. Together the protons and neutrons make up the *nucleus* of the atom. The nucleus has practically all of the mass of the atom.

The electron has a mass only about 1/1840 of that of the proton. It has a negative charge which is equal in value but opposite in polarity or sign to the positive charge of the proton. The quantity of charge e on an electron is the smallest charge that can exist. All charges are some integral multiple of this charge: 2e, 3e, etc. but never 1-1/4 e, 7-5/8 e, etc.

The neutron has no charge but a mass slightly more than that of a proton. Normally an atom will have as many electrons as it has protons, and it will be electrically neutral.

The orbits of the electrons form imaginary concentric shells around the nucleus with one or more electrons revolving in each shell. The electrons in the outermost shell, that is, the farthest from the nucleus, can break loose, under certain conditions, from the atom.

When electrons are thus removed from a body, the positively charged nuclei left behind will give the body a net positive charge. The body that receives these electrons will have an excess of electrons and will have a net negative charge. Now that we've looked at the atomic structure of matter, we can understand what happens when objects are electrified through rubbing or friction.

2.2 ELECTROSTATIC ATTRACTION AND REPULSION

When two electrified objects are brought close together, you will notice that they either attract each other or they repel. This can be readily demonstrated with a setup such as that shown in Fig. 2-1.

In this experiment a very light, small ball made of pith is suspended by a thread. A hard rubber rod is stroked with fur or flannel and then is made to touch the balls. After a brief instant, the ball will jump away from the rod as if it were being repelled.

Moving the rod near the ball again will still result in the ball moving away. But if now a glass rod is rubbed with silk and brought near the ball, the ball will be attracted to the rod. Once it contacts the rod, it will again be repelled by the rod.

In the process of rubbing, the rods acquired an electric charge. When they touched the pith ball, they transferred some of this charge to the ball. The ball then also became electrified.

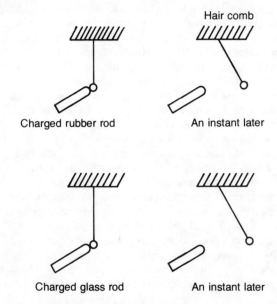

Fig. 2-1. Demonstration that like charges repel.

22

Because of this experiment and many similar ones, scientists have concluded that there are only two different kinds of electric charges. The charge on the rubber rod rubbed with flannel is called *negative* and the one on the glass rod, *positive,*

When the positively charged glass rod touched the ball, it gave it a positive charge and the ball was repelled by the rod. When the negatively charged rubber rod touched the ball, it gave it a negative charge and the ball was repelled by the rod. Thus it can be concluded that like charges repel one another.

If a positively charged pith ball is brought near a negatively charged pith ball, the two balls will attract each other. Thus, unlike charges attract one another.

It is often convenient to say that static electricity is created or generated. Actually, however, electric charge cannot be created; it is merely transferred. When the rubber rod in the experiment received a negative charge, the flannel received an equal and opposite positive charge. A *neutral* body has an equal amount of positive and negative charge; its net charge is zero.

As unlike charges attract and like charges repel, it is a simple matter to determine the charge on unknown bodies. Simply bring the body near a charged glass rod suspended by a thread and notice if the rod and the body repel each other. If so, the unknown body has a positive charge. If they attract, the unknown body has a negative charge.

The force between charges can be found by the use of Coulomb's Law. Charles Coulomb (1736–1806) was a French physicist who introduced the concept of *point charges,* which are charged bodies whose dimensions are small in comparison with the distance between them. He showed that the force F of attraction or repulsion between two point charges q_1 and q_2 is proportional to the product of the charges and inversely proportional to the square of the distance r between them. That is,

$$F = \frac{q_1 \, q_2}{k \, r^2}$$

where k is a constant called the dielectric constant of the medium. All mediums other than a vacuum act to lessen the force; however, the difference between air and a vacuum in this equation is negligible.

If the charges have like signs, the force will repel the charges; if the charges are opposite in sign, the force will attract the charges. Coulomb's law is *the* fundamental equation of electrostatics.

The name *coulomb* has been given to the unit of charge equivalent to the charge carried by 6.24×10^{18} electrons. As this is an enormous unit as far as electrostatics is concerned, smaller units such as the microcoulomb (μC), the nanocoulomb (nC), and the picocoulomb (pC) are used. Note that 1 μC = 10^{-6} C, 1 nC = 10^{-9} C, and 1 pC = 10^{-12} C.

2.3 TRIBOELECTRIC CHARGING

When movement occurs between two bodies, particularly if they are dissimilar, it will be found that one of the bodies will lose or give up electrons more readily than the other. In effect, electrons will be stripped or displaced from one body and transferred to the other. The body that lost electrons will have a positive charge while the body that received the electrons will have a negative charge.

The transfer of electrons takes place rapidly and then diminishes as the surface energies equilibrate. The generation of static electricity in this manner is called the *triboelectric effect*.

Each body will have an electrostatic potential that can be measured with an electrostatic voltmeter. The voltage generated may be 100 to 35,000 volts, as shown in Table 2-1. Its magnitude depends on speed of movement or separation, types of materials, humidity, surface characteristics, and surface geometry.

The movement that most often causes triboelectric charging is a simple rubbing together of the two bodies. In the process of rubbing, the bodies are brought repeatedly into close contact and then separated. Actually, rubbing is not necessary for triboelectric charging to occur. The act of separating two objects, even of the same material, that have been in contact can generate substantial electrostatic charges. This can be demonstrated by separating the sides of a plastic bag or unrolling transparent tape as shown in Fig. 2-2.

Rubbing or contact-and-separation can occur in many ordinary activities where no attempt is being made to generate static electricity. These actions include walking across a floor, taking

Table 2-1. Typical Electrostatic Voltages (DOD-HDBK 263).

Means of Static Generation	10 to 20 Percent Relative Humidity	65 to 90 Percent Relative Humidity
Walking across carpet	35,000	1,500
Walking over vinyl floor	12,000	250
Worker at bench	6,000	100
Vinyl envelopes for work instructions	7,000	600
Common poly bag picked up from bench	20,000	1,200
Work chair padded with polyurethane foam	18,000	1,500

24

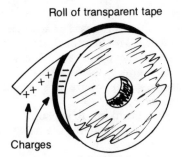

Roll of transparent tape

Charges

Fig. 2-2. Generation of static electricity by separating two objects.

off coats and sweaters, standing up from a chair, assembly line movement, removing an IC from a plastic envelope, and movement of ICs in DIP tubes during shipment. Table 2-2 lists other typical prime charge sources.

Table 2-2. Typical Prime Charge Sources (DOD-HDBK 263).

Object or Process	Material or Activity
Work Surfaces	Waxed, painted or varnished surfaces Common vinyl or plastics
Floors	Sealed concrete Waxed, finished wood Common vinyl tile or sheeting
Clothes	Common clean room smocks Common synthetic personnel garments Non-conductive shoes Virgin cotton*
Chairs	Finished wood Vinyl Fiberglass
Packaging and Handling	Common plastic—bags, wraps, envelopes Common bubble pack, foam Common plastic trays, plastic tote boxes, vials, parts bins
Assembly, Cleaning, Test and Repair Areas	Spray cleaners Common plastic solder suckers Solder irons with ungrounded tips Solvent brushes (synthetic bristles) Cleaning or drying by fluid or evaporation Temperature chambers Cryogenic sprays Heat guns and blowers Sand blasting Electrostatic copiers

* Virgin cotton can be a static source at low relative humidities such as below 30 percent.

25

The types of materials involved will help to determine the magnitude and polarity of the charges. These factors are related to the position of the materials in the triboelectric series, which is a list of substances in an order of positive to negative charging as a result of the triboelectric effect. A substance higher on the list, shown in Table 2-3, is positively charged when rubbed with a substance lower on the list.

The position, or order of ranking, of substances in this series should be considered as a rough guide, as the series is not always constant or repeatable. In general, the further apart two substances are in this series, the greater will be the magnitude of the charge created during triboelectric charging. It can be affected by surface cleanliness, humidity, lubricity, however, the amount of surface area involved in the rubbing action.

Positive +	Air
	Human Hands
	Asbestos
	Rabbit Fur
	Glass
	Mica
	Human Hair
	Nylon
	Wool
	Fur
	Lead
	Silk
	Aluminum
	Paper
	Cotton
	Steel
	Wood
	Amber
	Sealing Wax
	Hard Rubber
	Nickel, Copper
	Brass, Silver
	Gold, Platinum
	Sulfur
	Acetate Rayon
	Polyester
	Celluloid
	Orlon
	Polyurethane
	Polyethylene
	Polypropylene
	PVC (Vinyl)
	KEL F
	Silicon
Negative −	Teflon

Table 2-3. Triboelectric Series (DOD-HDBK 263).

Notice in Table 2-3 that cotton is at the center because it is relatively neutral. The materials above it in the table are increasingly positive; those beneath it, increasingly negative. A material can have either a positive or negative charge, depending on what it is rubbed with.

The larger the contact area between the two materials, the more electrons that can be involved in the transfer. Thus, the magnitude of the triboelectric charge is proportional to the contact area. By minimizing the contact area through the use of ribs or standoffs, the charging of such devices as ICs in shipping tubes can be reduced.[1]

Because triboelectric charging is a friction process, one way to reduce it is to increase the material's lubricity, which is a measure of surface smoothness and frictional characteristics. The higher the lubricity of the surfaces being rubbed, the lower the friction and hence the lower the generated charges. Lubricity can be increased through the use of materials called *antistats*.[2]

The charges developed during triboelectric charging are either mobile or immobile. The charges on conductors are *mobile* and are rapidly distributed over their surfaces and the surfaces of other conductive objects they touch. The charges on insulators are *immobile;* they tend to remain in the localized area of contact. Since immobile charges are not readily distributed over the entire surface of the substance, they generate high electrostatic voltage levels on insulators.

Electrostatic charges can be readily transmitted from an object to a person's conductive sweat layer causing that person to be charged. When a charged person handles or comes in close proximity to an electrostatic sensitive part, he can damage that part either by direct discharge (by touching the part) or by subjecting the part to an electrostatic field.

One of the basic laws of nature is that the net electric charge of any isolated (insulated) system remains constant. Charges can only be separated or combined; they cannot be created. Thus, in triboelectric charging, when charges are transferred from one body to another, for every negative charge there will be an equal positive charge. What one body gains, another loses.

2.4 CONDUCTORS AND INSULATORS

In some of the earliest experiments with electrostatics, it was found that charges could be easily transferred from one point to another by means of wires. Thus, metals, whatever their shape or configuration, are considered to be good *conductors* of electrical charges. They have high *conductivity,* the ability to readily pass a current, a flow of charges or electrons.

Within solid conductors of electricity there are millions of free electrons which can move or wander among the atoms. That is, these electrons can be temporarily detached from the atoms. When a conductor is attached or connected to a charged body and to ground, the free electrons will flow in a continuous stream, in a definite direction, and thereby transfer the charge. As the electrons have like negative charges, this movement is a result of the force of repulsion between the electrons and the force of attraction between the electrons and the positive charges.

In other substances such as rubber, glass, and mica there are very few free electrons and consequently great opposition to electric currents. Because of this opposition, these substances have very low conductivity and thereby are classified as non-conductors or *insulators*.

The opposition to the flow of electrons is called *resistance* and is measured in ohms. The resistance (R) of a piece of material is inversely proportional to the cross sectional area (A) perpendicular to the flow of current and directly proportional to the length of the material parallel to the flow of the current. That is, the thicker a conductor, the less will be its resistance but the longer it is, the greater will be its resistance. This can be expressed as:

$$R = \rho_v \frac{\ell}{A}$$

where ρ_v is a constant known as *volume resistivity*. Volume resistivity is expressed in ohms per cm and is a constant for a given homogeneous material.[3]

The equation may be rewritten as

$$\rho_v = \frac{RA}{\ell}$$

Thus, the volume resistivity of a homogeneous material can be determined by measuring the resistance of a piece of the material with known dimensions (that is, length ℓ, width w, and thickness t).

For a square piece of material, ℓ is equal to w and the equation can be simplified to

$$\rho_v = R\,t$$

Thus, ρ_v is normally determined by measuring the resistance R of a square of material and multiplying it by the thickness t.

Volume resistivity is an inverse measure of the conductivity of a material. In an electrical insulating material, it is numerically equal to the volume resistance in ohms between opposite faces of a 1 cm cube of the material.

While volume resistivity is the primary unit of measurement, surface resistivity (or sheet resistance) is also frequently used. Originally it was a convenient term applied to thin films of materials, generally metallization layers. It is now used by industry to describe the conductivity of much thicker materials.[4]

Surface resistivity is numerically equal to the surface resistance (R) of a square section of material of a given thickness. (the size of the square is irrelevant). It is measured in ohms per square.

Surface resistivity is commonly used as a resistance measurement parameter of laminated materials having a thin conductive surface over an insulative base. It is used to define the resistivity of surface conductive materials such as hygroscopic anti-static polyethylenes, plastics, and other conductively coated or laminated insulative materials.[3]

Conductive layers on these materials usually have near uniform thickness such as the sweat layer of hygroscopic anti-static materials. The surface resistivity does not effectively change by increasing or decreasing the thickness of the base insulative material if its volume resistivity is high in relation to that of the conductive surface materials.[3] Per military specifications, surface resistivity measurements are used to describe three basic types of materials used for ESD protection: conductive, static dissipative, and anti-static.

Conductive materials have surface resistivities of 10^5 ohms per square or less. Metals, some bulk conductive plastics, wire impregnated materials, and laminates can meet this requirement.

A bulk conductive material with a volume resistivity of 10^4 ohms per cm would be conductive, if its thickness was 0.1 cm or greater. Such a material with a thickness of less than 0.1 cm, however, would be classified as static dissipative.

Static dissipative materials have surface resistivities of greater than 10^5 but less than 10^9 ohms per square. Some materials normally considered conductive can be made very thin and have a surface resistivity between 10^5 and 10^9 ohms per square placing them in the static dissipative category.

Anti-static materials have surface resistivities equal to or greater than 10^9 but less than 10^{14} ohms per square. (Some authorities give 10^{13} as the upper limit.) These materials include hygroscopic anti-static materials such as some melamine laminates, high resistance bulk conductive plastics, wood and paper products, and very thin layers or films of static dissipative or conductive materials.

In many anti-static materials, hygroscopic (water seeking) agents are added to the plastic during manufacture. These agents constantly migrate to the surface where they attract atmospheric moisture. The resulting monolayer of water forms a thin electrically conducting surface which dissipates static charges.[5]

In conductive plastics, a mixture of carbon powders (which are conductors) is added to produce an electrically conductive material. To produce an effective conductive path to ground it may be necessary to add up to 40% by volume of the carbon to the plastic. This addition unfortunately may mechanically weaken the plastic, allowing it to tear and puncture more easily.[6]

Another technique for lowering the resistivity of plastics is to add metallic fibers, forming metalloplastics. In a patented process called Cross-Link™ resistivities as low as 0.001 ohms per cm have been obtained for such metalloplastics. With Cross-Link it is theoretically possible to make any plastic conductive with as little as 1% fiber by volume.[7]

Still another technique to accomplish static dissipation is used in Benstat™ materials. In these materials the chemical additive amounts to only 10% of the material. This additive is reportedly chemically bonded with the molecular structure of the base polymer to produce an ionic path through the entire volume of the material. Such materials are said to be permanently antistatic and not dependent on moisture for conductivity.[6]

Once a charge is generated, its distribution depends on the resistivity and surface area of the material. That is, the more conductive the material, the faster the charge will be distributed. The greater the surface area over which a charge is spread, the lower will be the charge density and the level of the residual voltage. As we have noted earlier, in contrast to insulators, localized charges cannot exist on conductors.[3]

Because of this effect, conductive bodies and materials are used for electrostatic discharge control. But there is a distinct limitation with such materials. If, for example, a charged circuit pack or semiconductor approaches a highly conductive object, say a tote box or table top, a spark and high discharge current could occur. When this occurs, any semiconductors in the discharge path may be damaged. To prevent this, the tote box or table top should be made conductive enough so that significant voltages will not be induced across the tote box or table top, but not be so conductive that a discharge will rapidly occur.[3]

When an object with an electrostatic charge is placed on a static dissipative or conductive surface, the charge will gradually dissipate or decay. Although this decay may appear to be instantaneous, it may take several hundredths of a second up to several seconds.

Decay time generally is measured by charging a section of material with a static voltage and measuring the time for the voltage to decay to a given level such as 10% of its original value. The Electronic Industries Association (EIA) specifies that bags and pouches for carrying electrostatic sensitive components must have an electrostatic decay time of not more than 2.0 seconds when measuring both an applied positive and negative 5000 volt charge, dissipating to 50 volts after the material is grounded. In general, decay time is related to conductivity.

Until now we have considered an electric current to be a flow of electrons. However, in gases and liquids the flow of electricity is a movement of *ions*, rather than of electrons. As you may recall, if an atom or molecule loses one or more electrons, it becomes a *positive* ion; if it obtains more than its normal share of electrons, it becomes a *negative* ion.

Ordinarily air is a poor conductor because most of its molecules are electrically neutral. Some of the molecules will be found to be ionized because of ultraviolet light or cosmic rays. When two strong opposite charges are brought close together, the force between them may give enough energy to these random ions that they will collide with other molecules, dislodging electrons and thereby creating additonal ions. In a brief instant, a chain reaction occurs, making the air a conductor and allowing a sudden rush or discharge of electrons to bridge the gap between the two bodies. The visible flash that results is called an electric spark. The transfer of charge by ions can also take place in liquids as well as in gases.

From a chemical point of view, rather than biological, water is said to be pure if it has no dissolved minerals. From a practical view, such water is a very poor conductor. But of course pure water is not at all common; it is likely to be found only in a laboratory. Ordinary drinking water, surprisingly, is a fair conductor because of the dissolved impurities.

Even atmospheric moisture is likely to have impurities. This conclusion follows from the long observed phenomenon that static electricity is more noticeable (and troublesome) on dry days than on damp days. In fact, when the relative humidity is very high, it becomes most difficult to perform classroom demonstrations of generation of static electricity. On damp days, the moisture condenses on many objects. While not necessarily visible, it forms a very thin coat or film of water which allows charges to dissipate.

Movement of ions in films of water and air may be a better

explanation of some triboelectric charging than is transfer of electrons. In the classic experiment of rubbing a rubber rod with fur, we assume these objects become charged because electrons have been moved from one object to the other. But since rubber and fur are poor conductors, they would not be expected to have any significant number of free electrons. The charge generated may be more likely due to movement of ions.[8]

Another good conductor is the earth itself unless it is extremely dry or rocky or both in the place where the resistance is measured. If a charged object touches or is connected to ground, it is said to be *grounded*. It is not necessary in all cases for the connection to ground to be made by a metallic conductor. The human body may be a conductor to ground, although not a very good one.

The resistance of the human body can vary from 100 to 100,000 ohms, depending on the amount of mositure, salt and oils at the skin surface, skin contact area, and pressure. Typically it is between 1,000 and 5,000 ohms for actions which are considered pertinent to holding or touching parts or containers such as finger-thumb grasp, hand holding, or palm touch. A value of 1,500 ohms provides a reasonable lower human resistance value. As we shall see later in this chapter, this value has been picked to represent the human body in standard test circuits.

Because the human body can have a low resistance, common voltages present in some assembly and test procedures may be lethal if a person accidentally touches an electrically live circuit and ground at the same time. To minimize such risks it is common practice in electrostatic protected work stations to use *soft* grounds. A soft ground is a connection to ground through a resistor sufficiently high to limit current flow to a safe level, normally 5 milliamperes, for personnel. The value of the resistor depends upon the voltage levels which could be contacted by personnel who were so grounded.

In the most common application of a soft ground, a resistor is placed in series with a person's wrist strap and the ground connection. If there were no resistor in this situation, if the wrist strap had a direct connection to ground, then a *hard* ground would exist.

2.5 CHARGING BY INDUCTION

Charging by *conduction* is probably the most common method whereby a conductor becomes electrified. But charging by *induction,* where there is no physical contact between the charging source and the object to be charged, is also a significant method of charging an object. It can be demonstrated using the equipment shown in Fig. 2-3.

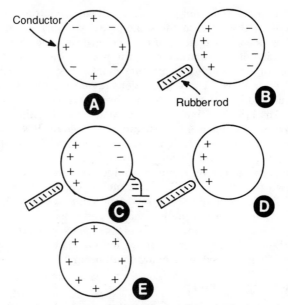

Fig. 2-3. Charging by induction. (A) Isolated, uncharged conductor. (B) Conductor charged by rubber rod. (C) Ground connection allows electrons to flow to ground. (D) With ground connection removed, conductor has only positive charges. (E) With rod removed, charges distribute evenly.

When a negatively charged rubber rod, for example, is brought near an insulated metal conductor, the electrons in the rod will force many of the electrons in the conductor to the opposite side of the conductor. The conductor's shape incidentally is irrelevant; it may be a sphere, rod, plate, or whatever.

Now while the rod is still in the vicinity of the conductor, a person touches any part of the conductor or connects a wire from the conductor to ground. The electrons in the conductor will flow through the person or the wire to the ground as they are being forced away from the rod. After a brief instant, the ground connection is broken. With no free electrons left on the conductor, the conductor now has a net positive charge. The positive charges will distribute themselves evenly over the conductor when the rod is removed from the vicinity.

If we had used a positively charged rod, it would have attracted negative charges (free electrons) to the near end of the conductor, leaving the far end with a positive charge. By grounding the conductor, electrons are pulled up into the conductor from the earth, giving the conductor a net negative charge.

Thus, whenever a charged object is brought near a conductor

that is insulated from ground, it will cause charge separation, leaving a charge with the same polarity (sign) at the opposite end or side of the conductor and a charge of opposite sign at the near side of the conductor. In the process the charging object does not lose or give up any of its charge. Just how much charge will be induced on an object depends on the object's size and shape and how near it is to the charging body.

An investigation of integrated circuit shipping tubes showed that charges were being developed on ICs due to their movement in the tubes. This charge consisted of a mobile charge on the metal lead frame and conductors of the IC and an immobile charge on the non-conductive portions of the IC, as shown in Fig. 2-4. While the immobile charge cannot develop currents (and therefore cannot directly damage the IC), it can by induction produce a mobile charge and electrostatic discharge on the lead frame of the IC. This takes place when the immobile charge induces a charge separation on the frame. Then, if the frame is grounded, a discharge will occur.[9]

2.6 ELECTRIC FIELDS

In previous paragraphs we've noticed that electrically charged bodies can attract or repel each other through a distance even though there is absolutely no substance, no gas, no liquid between them. It can occur even in a vacuum. How this action at

UNCHARGED DIP

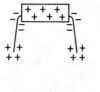

DIP WITH IMMOBILE CHARGE ON BODY INDUCES CHARGE SEPARATION ON LEAD FRAME.

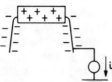

ESD RESULTS WHEN GROUNDING LEAD FRAME. DEVICE LEAD FRAME IS NOW CHARGED.

Fig. 2-4. Charge induced on device lead due to immobile charge on package body (courtesy of Burt Unger[9]).

a distance occurs cannot be precisely explained and indeed it's not necessary to have a mental picture of what is happening in order to perform calculations in electrostatics. Scientists, nevertheless, have devised a model which makes this concept easier to understand. It has two distinguishing features: a force field and lines of force.

The space or region in the vicinity of an electric charge is not the same as an identical space which has no charge in it. An object inserted in this space will have different properties (force, charge, potential) than it would have in a space with no nearby charge. The charge is said to alter the space although there is nothing there to be changed, at least in the case of a vacuum. The altered space is called an *electric field.*

At any point in this field an electric charge, q, will encounter either a force of attraction or a force of repulsion. The force, F, is proportional to the charge, q, and the strength, E, or intensity of the electric field. That is,

Force = charge X strength of electric field

or $F = q E$ where E and F are vectors.

If q is positive, the force will be in the direction of the field (a concept we will explain shortly). If q is negative, the force will be opposite to the direction of the field.

The electric field strength, E, at any point is the force, F, on a small positive test charge, q_1, divided by q. That is,

$$E = F/q_1$$

In this calculation it is assumed that the test charge is small enough not to upset the field, that is, to change the charge distribution and thereby give a false indication. For most practical purposes the electric field has influence in the space immediately adjacent to the charge that created the field. In theory, however, the field extends to infinity. As you might expect, its strength is negligible at distances beyond the immediate vicinity.

The equation $E = F/q$ can be combined with Coulomb's law to show that the electric field strength at a point that is at a distance, r, from the charge is

$$E = k \frac{q}{r^2}$$

where k is a constant.

There is no electric field inside a solid conductor provided there is no current through the conductor. If there were an

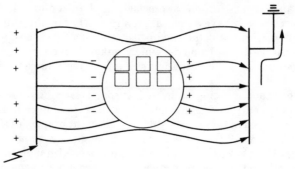

Fig. 2-5. There is no electric field inside a Faraday cage (courtesy of S. Y. Youn[10]).

electric field in such a conductor, it would cause the free electrons to move; but electrostatics is the science of charges at rest.

Faraday showed that there is no electric field inside a closed metal container as shown in Fig. 2-5. A Faraday cage is an electrically continuous, conductive enclosure which is usually grounded although it need not be. It is a widely used device for electrostatic shielding since an enclosed object is protected from external fields and discharges. If there is no electric field within a conductor, it follows that all of the charge on the conductor lies on its surface.

If instead of an isolated charge, there are a number of individual charges in the region, the electric field gets complex. Its intensity at any given point will be the vector sum of the electric field from all these charges.

The configuration of an electric field can be shown with the use of imaginary *lines of force,* which give the direction of the electric intensity. As shown in Fig. 2-6, these can be straight lines or curves. The arrow on each line indicates the direction of a small positive charge would take if placed at the point. (A negative charge would travel in the opposite direction.)

Note that the lines are in three dimensions, not just the two shown in the illustration. The lines from a point charge, for example, bristle out in all directions. Because of space limitations on the illustration, it may not be obvious that all of these lines are continuous between a positive charge and a negative charge. That is, each line must start on a positive charge and end on a negative charge. For isolated charges, the opposite charge is imaged at infinity. That is, for isolated charges the lines extend to infinity as shown in Fig. 2-6A.

For an isolated negative charge, the arrows would point in the opposite direction of that shown in the illustration, indicating that the lines started at infinity and ended at the charge. To show that a field is weaker in some directions, the lines are drawn farther apart. However close they are drawn, the lines never cross.

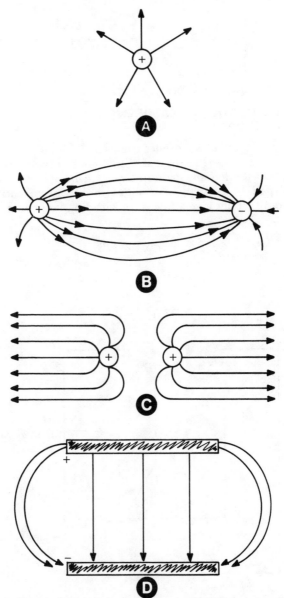

Fig. 2-6. Electric fields. (A) Isolated charge. (B) Unlike charges (electric dipole). (C) Like charges. (D) Parallel flat plates.

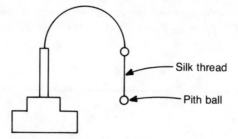

Fig. 2-7. Pith ball electroscope.

2.7 ELECTROSTATIC MEASUREMENTS: THE ELECTROSCOPE

A small pith ball suspended from a metal stand, as shown in Fig. 2-7, can be used as an instrument to *detect* static charges. Notice that it allows no quantitative *measure* of the charge. As the pith ball is very light, it can be disturbed by air currents in the room and thereby give a false or misleading indication. While

Shown in charged position.

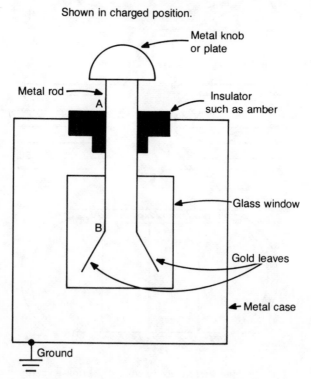

Fig. 2-8. Gold-leaf electroscope.

such a device is called an "electroscope," the term "electroscope" is most commonly used to mean "gold-leaf electroscope."

The electroscope is a much more sensitive instrument which can be calibrated to indicate the magnitude of a charge. An early device (it's about 200 years old), it helped scientists formulate the law of electrostatics.

It consists of two very light thin strips (or "leaves") of gold or aluminum suspended side by side, as shown in Fig. 2-8, from a metal rod (AB). When uncharged, the leaves hang loosely. The surrounding case or box which is either glass or metal with a glass window for observation, protects the gold leaves from air currents. If the case is all glass, the insulated bushing or collar is not needed. When a metal case is used, it is generally connected to a water pipe or other electrical ground via a copper wire. Once grounded the metal case cannot become charged and thereby disturb the leaves.

When the metal ball on top of the electroscope touches a positively charged body, as shown in Fig. 2-9, the positively charged body will attract the electrons from the leaves, rod, and sphere. As a result the leaves are left with a positive charge. Because the leaves have like charges, they will fly apart from the force of repulsion. The greater the charge on these leaves, the farther the leaves will separate, that is, the greater will be the angle of divergence.

While this example has shown a positively charged body, the electroscope works equally well for negatively charged bodies, as indicated in Fig. 2-10. Note that the negatively charged body forces electrons into the leaves which then separate because they have like charges.

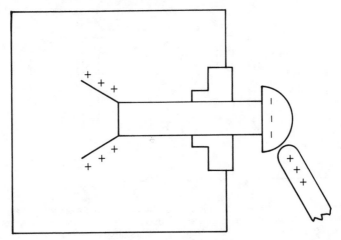

Fig. 2-9. Charging electroscope with positive source.

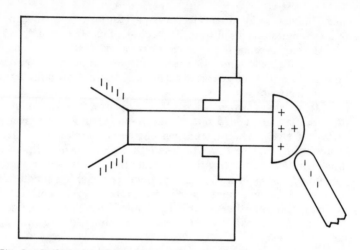

Fig. 2-10. Charging electroscope with negative source.

Whether positive or negative, the charge to be detected and/or measured in effect immediately distributes itself over the leaves. That is, the charge on the leaves will have the same sign (positive or negative) as the object.

If the sign of the charge on the electroscope is known, then the electroscope can be used to find the sign of an unknown charge. For example, if the electroscope is positively charged, as shown in Fig. 2-11A, then if a charged rod touches the electroscope, we can determine its polarity by the reaction of the leaves in the electroscope. If the leaves spread even farther apart, the

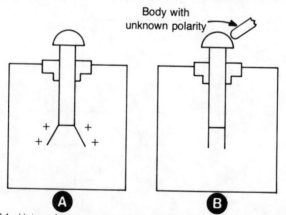

Fig. 2-11. Using electroscope to determine polarity of a body. (A) Electroscope with positive charge. (B) Electroscope after being touched by body with unknown polarity (obviously a negative polarity to drain the positive charge in (A).

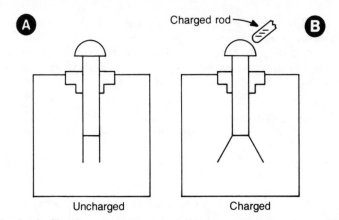

Fig. 2-12. Charging electroscope by induction.

rod has a positive charge; if the leaves pull together, the rod has a negative charge (Fig. 2-11B).

In some cases the charge on an object can be so great that it may damage the electroscope if the object touches the knob on the electroscope. But it is not necessary to touch the object whose charge is to be measured. Through induction if the body is simply brought near the electroscope, the electroscope will detect the presence of a charge. This is shown in Fig. 2-12.

To change the electroscope from a charge measuring device to a potential measuring device, that is, to make it into a simple voltmeter, it is necessary only to add a scale to measure the angle of deflection of the leaves. In this procedure, the sphere on the electroscope is connected to the body whose potential is to be measured by a long wire.

2.8 ELECTRIC POTENTIAL

If a small positive test charge, q, is placed in an electric field associated with charge Q, as shown in Fig. 2-13, it will require work to push or move it from point A to point B, as energy must be expended in overcoming the force of repulsion. This work is stored as potential energy in q. (To move the test charge from

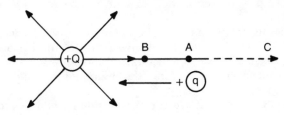

Fig. 2-13. Electric potential.

41

point B to A, the electric field will do work and the potential energy of q will decrease.)

The work, W, done in moving this unit positive charge is defined as the *potential difference* V between the two points. The three quantities are related by the equation

$$\text{Potential difference} = \frac{\text{work done}}{\text{quantity of charge transferred}}$$

Or
$$V = \frac{W}{q}$$

When q is in coulombs and W is in joules, then V is expressed in *volts*, a practical unit named in honor of the Italian physicist Volta (1745–1827). Thus, if 1 joule of work is required to move 1 coulomb of charge from point A to point B, the potential difference between the two points is 1 volt.

Because potential difference is a scalar quantity, it can be added to other potential differences algebraically. The potential difference between two points is also referred to as the *voltage* between the two points.

In Fig. 2-13 the path of q from A to B is a straight line. However, even if the path had been curved and haphazard, the work required would have been the same. A positive charge will tend to move from a point of high potential to a point of lower potential, that is, from B to A in Fig. 2-13. Point B will be at a higher potential than point A because it is closer to the source.

The line of force always points from higher to lower potentials for positive charges. Note that point B may have, for example, a potential of 120 volts as referenced to earth and point A a potential of 100 volts. The potential difference between A and B is the simple difference of 20 volts. To move a positive charge from a point of lower potential to a higher potential will require work; the converse is true for negative charges.

In making measurements of difference of potential, it is often convenient to determine the voltage between a point and the earth (ground). In this case, the potential of the earth is considered to be zero. Another point considered to be zero is infinity.

Because of the earth's zero potential, any conductor connected to it will discharge and have a zero potential also. By the same token, any two conductors will have the same potential when connected together, once the circuit has settled down and current no longer flows.

While there is no electric field inside a hollow conductor, such as a sphere, there can be a region of uniform potential in it.

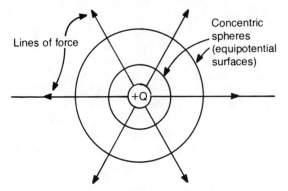

Fig. 2-14. Equipotential surface.

The potential on the inside wall will be the same as on the outside.

2.9 EQUIPOTENTIAL SURFACES

Within an electric field there will be numerous points with the same potential. If we draw an imaginary surface to connect these points, we have an *equipotential surface*. That is, all points on this surface have the same potential. Because there is no potential difference between any two points on the surface, no work is required to move a charge around on it.

When the surface is drawn, it will be found to be at right angles to the electric field. The surface may be a plane but generally is some curved shape. In the case of an isolated point charge, the equipotential surfaces will form concentric spheres around the charge, as shown in Fig. 2-14. Only two spheres are shown, but an infinite number could be drawn.

Figure 2-15 shows the electric force on charge q, which is moving at right angles to the line of force. No work is required to move q from A to C, a very small distance. If q keeps moving at right angles to each line of force it encounters, it will map out an equipotential surface.

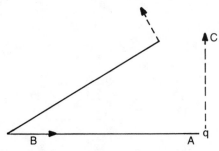

Fig. 2-15. Charge movement without work.

While no work would be done in moving a charge on an equipotential surface, work would be necessary to move a charge from one equipotential surface to another, as from the outer sphere in Fig. 2-14 to the inner sphere.

As we have noted earlier, a static electric field cannot exist within a conductor when the charges in the conductor are at rest. As a consequence of this, in electrostatics the surface of a single conducting body is an equipotential surface. That is, there will be no difference of potential between any two points of a conductor. Here again the electric field near the surface will be perpendicular to the conductor.

2.10 SURFACE CHARGE DENSITY

Electrostatic charges exist only on the outside or surface of a conductor, whether solid or hollow, never inside. The charges are distributed uniformly over spherical conductors; over other shapes, such as egg-shaped conductors, the charges are not distributed evenly. As shown in Fig. 2-16, they are greatest where the surface has the most curvature, that is, the ends or projections. The sharper the curve, the greater will be the charge density and therefore the greater the chances for the charges to escape. The charges escape as the surrounding air becomes ionized and corona discharge occurs.

Corona discharge is particularly noticeable with pointed surfaces such as needle shape. When Ben Franklin discovered this phenomenon he put it to use in the lightning rod. In general, a grounded sharp metallic point can be used to discharge charged conductors without sparking simply by bringing the point near the conductor.[11]

If electric discharge is not desired, pointed surfaces or protusions should be avoided. Instead, spherical surfaces with a high radius of curvature should be used. The larger the diameter of the sphere, the greater will be the potential on it before the surrounding air breaks down and conducts. This principle is used in Van de Graaff generators, which will be discussed in later paragraphs.

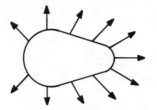

Fig. 2-16. Surface charge density.

2.11 CAPACITANCE

When two conductors are in close proximity but not touching, they form a *capacitor*. Whether this condition is deliberate or accidental, a capacitor can readily store electric charges when connected to a source of potential difference such as an alternating current generator or a battery. When so connected, one conductor will acquire a positive charge, the other a negative charge. The charge, Q, of the capacitor is considered to be the charge on either conductor, as the charges are equal but opposite.

Because the charges are separated by an insulator such as air or mica, they cannot flow through the capacitor. Instead, they pile up on the conductors. There they stay until either the potential is removed, the polarity is reversed, or a wire is connected between the two conductors.

In the first case, the charges eventually dissipate. In the second case the positive and negative charges effectively change places. In the third case, the energy is suddenly released in a discharge initially occurring as a spark.

The ability of a capacitor to store a charge is called *capacitance*. Capacitance, C, is directly proportional to the charge, Q, on either conductor and inversely proportional to the voltage (potential difference) between the conductors. That is, $C = Q/V$ where C is measured in farads, Q in coulombs, and V in volts.

The unit farad was named in honor of Michael Faraday. As it is a very large unit, fractional units of microfarad (μF) and picofarad (pF) are more commonly used. In conventional electronic equipment, capacitors will have values that typically range from 1 pF (1 pF = 10^{-12} farad) to thousands of microfarads (1μF = 10^{-6} farad).

Capacitors may be constructed in a variety of forms: concentric spheres, concentric (coaxial) cylinders, or, the most common, parallel plates. The first two are found only in laboratories; the latter, however, is used in millions of radio and television sets and other electronic gear. Indeed, it is hard to find an electronic device that does not contain at least one capacitor.

The parallel plate capacitor has three common configurations: (1) small metal plates that intermesh, when turned on a common rotor, with other metal plates without touching; the plates are separated by air; (2) stacks of fixed position metal plates separated by small sheets of mica, ceramic, or other insulators; and (3) long thin strips of metal foil separated by a nonconducting material (dielectric) such as waxed paper. In the latter, the foil and paper are rolled together into the form of a cylinder.

Just how much capacitance a given capacitor has depends on the distance between the plates, the nature of the dielectric separating the plates, and the size of the plates. The capacitance increases when the plates are brought closer together. When the plates are pulled apart, the capacitance decreases. The greater the area of the plate, the higher will be the capacitance.

The capacitors discussed thus far are *discrete* capacitors and are used as circuit elements. However, all conductors form a capacitor to their neighboring conductors, that is, paths on a printed cicuit board, or to ground with the air space between being the dielectric. These are generally referred to as *parasitic* capacitors and have low values, generally in the picofarad range.

The metallic parts of a semiconductor device such as a DIP can act as one plate of a capacitor with the ground plane forming the other plate. As shown in Fig. 2-17, the capacitance will vary with respect to the device's orientation and proximity to ground.[1]

2.11.1 Dielectrics

When a solid dielectric is inserted between the air-separated plates of a charged capacitor, the electric intensity will be weakened and therefore the potential difference (V) between the plates will decrease. In effect, for a given fixed charge, when the potential is decreased, the capacitance will be increased as can be seen from the equation $C = Q/V$.

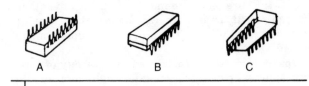

Orientation	Device capacitance (pF)				
	16 Pin DIP (Plastic)	18 Pin DIP (Plastic)	24 Pin DIP (Plastic)	24 Pin DIP (Ceramic side-brazed)	40 Pin DIP
A	2.9	3.6	7.1	28	52
B	2.0	2.3	3.9	3.6	6.6
C	1.4	1.6	2.0	2.0	2.8
$\dfrac{C_{MAX}}{C_{MIN}}$	2.0	2.2	3.5	14	18.6

Fig. 2-17. Capacitance of device packages with respect to orientation to a ground plane. The ground plane is separated from the device by a thin insulator. (courtesy of P. R. Bossard[1]).

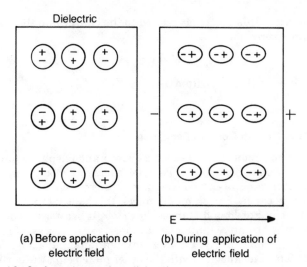

(a) Before application of
electric field

(b) During application of
electric field

Fig. 2-18. Surface charges in a dielectric.

How much larger the capacitance will become depends directly on the type of material used as a dielectric. This factor is called the *dielectric constant,* which is a simple number. The higher this constant, the greater will be the capacitance of the capacitor in which this dielectric is used. Representative values are 1 for air, 1 to 3 for many common plastics, 4.5 to 7.5 for mica, and 4.8 to 10 for glass and ceramics. If the solid dielectric is removed, the potential on the capacitor will return to its value before the dielectric was inserted.

What causes this dielectric effect? In a nonconductor, electrons cannot leave the atom or molecule to which they belong. However, in the presence of an electric field, they may shift their position so as to get closer to the positive electric field. When this shift occurs, the electrical center of charge of the electrons is no longer the nuclei of the molecule. The molecule becomes an electric dipole and the material is said to be polarized. These molecules line up as shown in Fig. 2-18. Within the interior of the material, the plus and minus charges cancel each other so that there is no net charge inside the material.

But notice the dipoles on each side of the material. As there is no matching charge, the left side, in this case, will have a negative charge and the right side a positive charge. The charges on these surfaces are called *bound* charges. These bound charges reduce the electric field intensity as they are in opposition to it.

2.11.2 Energy Stored in a Capacitor

When a capacitor is charged, work must be done to transfer the charges from one plate to the other. The total work, W, involved in this process is related to the charge and the voltage by the expression

$$W = 1/2 \, QV$$

which may also be written as

$$W = 1/2 \, CV^2$$

2.11.3 Human Body Capacitance

Some circuit elements such as power and telephone lines are *unintentional* capacitors. That is, there may be appreciable capacitance between two conductors or between one conductor and the earth. For short distances this capacitance may be negligible. For long distances, however, it becomes significant and must be entered into calculations for line losses, etc.

The human body, another unintentional capacitor, forms one plate of a capacitor with the other plate being the earth. Its capacitance depends on the amount and type of clothing and footwear, differences in floor materials, and the person's proximity to ground. Although it may be as high as several thousand picofarads, typically it ranges from 50 to 250 pF which is a very small value. Tests have shown that 80% of the population has a capacitance of 100 pF or less. However small, it nevertheless is enough to store sufficient charge to cause extensive ESD damage.

2.11.4 Capacitance Effects With Respect to Ground

The equation for capacitance can be rewritten as $Q = CV$, where, as before, Q is the charge, C the capacitance, and V the voltage. If Q does not change, then if C decreases, V must increase. That is, if the capacitance is continually decreased, the voltage will increase until a discharge occurs via an arc. For example, the charge potential on common polyethylene bags may jump from a few hundred volts while lying on a bench to several thousand volts when a person picks them up, because of the decrease in capacitance.[3]

2.12 ELECTROSTATIC GENERATORS

While great efforts are made in many industries to prevent the generation of static electricity, there are occasions when it is desirable to generate static electricity under controlled conditions for very specific purposes. The most common devices for these uses are the Van de Graaff generator and the human body ESD simulator.

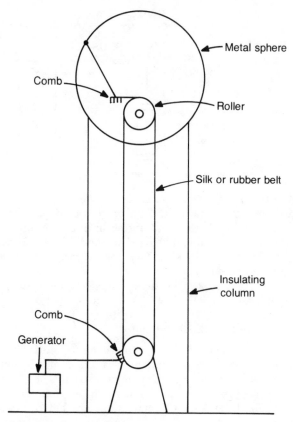

Fig. 2-19. Van de Graaff electrostatic generator.

2.12.1 Van de Graaff Generator

A common laboratory device for generating electrostatic charges continuously is the Van de Graaff generator, shown in Fig. 2-19. The smaller versions of this generator are used in classroom demonstrations while the larger ones, which are several feet tall, are used in nuclear physics experiments and in the production of high intensity X-rays. In the latter version potentials of 6 million volts have been generated.

As shown in the illustration, a motor-driven silk or rubber belt carries charges from comb A to comb B where they are transferred to the interior of the metal shell. The charges remain on the interior only momentarily before being forced to the exterior because electrostatic charges always reside on the exterior of a conductor.

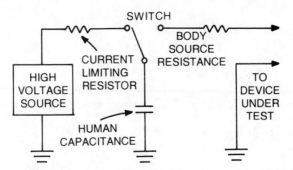

Fig. 2-20. Human body model ESD simulator (courtesy of Jack E. Keller[13]).

With the interior free of charges, it is ready to receive more charges coming up the belt. This process continues with the potential becoming higher and higher until the charges on the sphere become great enough to repel any additional charges or to start leaking off into the air. The potential can be increased by making the sphere larger.

In manufacturing operations, conveyor belts sometimes become unintentional Van de Graaff generators.[12]

2.12.2 Human Body ESD Simulator

Because people are a prime source of electrostatic discharge for damaging parts, an equivalent circuit is needed to simulate the charge storage and discharge characteristics of the human body. Various types of electrostatic discharge simulators have been used in ESD tests of semiconductors but the industry

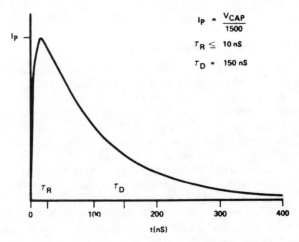

Fig. 2-21. Static discharge short circuit current waveform (courtesy of R. J. Antinone[15]).

standard is now a resistor-capacitor network. This circuit has the advantage that its action can be easily and consistently repeated.

As shown in Fig. 2-20, the simulator has a current limiting resistor to control charging of a capacitor. The human body capacitance is the sum of two capacitances: (1) that due to the body's isolation from ground when the body is considered to be a sphere with the same surface area and (2) that due to the parallel plate capacitive effect of the human body in contact with a ground plane through the soles of the shoes.[13]

The switch shown in this simulator is generally a bounceless high-voltage relay. It is shown in the charge position. At least one researcher believes that the high inherent capacitance of this relay switch must be considered in the design.[14]

Industry-wide values for the current limiting resistor range from 470 kilohms to 25 megohms, the body source resistance from 300 ohms to 4 kilohms, and the voltage source from 100 to 15,000 volts. The simulator specified by DOD Handbook 263 gives the value of capacitance as 100 pF $\pm$ 5%, the body source resistance as a noninductive 1.5 kilohms $\pm$ 5%, and the high voltage as variable between 0 and 15,000 volts dc. The waveform of the static discharge produced by such a simulator is shown in Fig. 2-21.

Commercial electrostatic discharge simulators typically include a power supply, capacitor, resistance, and meters, all with variable values in addition to the standards set by DOD Handbook 263.

Whether the values set by DOD Handbook 263 are the worst case or not is a matter of controversy. Some feel that under more stringent human model conditions, a part could be sensitive to ESD even though tests with the simulator indicated that the part was not ESD sensitive. Others,[4] however, feel that 1.5 kilohms and 100 pF do give a worst case. In real life the discharge contact is always accompanied with corona or a predischarge that tends to diminish the 100 pF energy effects on the device. In fact, a

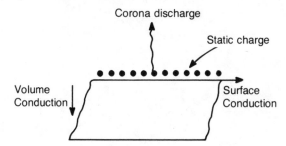

Fig. 2-22. Static charge dissipation (courtesy of George R. Berbeco[17]).

device that fails a test at 500 volts can probably take 2 to 3 times the voltage in real life situations.

2.13 STATIC CHARGE DISSIPATION

As shown in Fig. 2-22, a static charge can be dissipated by surface or volume conduction to ground or by corona discharge into the air. The better these paths of conduction, the more effective will be the dissipation. Corona discharge can be increased through ionization of the surrounding air. Surface and volume conduction can be increased by using materials with lower surface and volume resistivities.[16]

3

Electrostatic Test Equipment

W HILE STATIC DISCHARGES CAN SOMETIMES BE SEEN OR FELT, TEST equipment is necessary to reliably locate and measure electrostatic problems. The instruments described in this chapter are necessary for a full scale electrostatic protection program. For the hobbyist or small electronics shop, such equipment obviously is prohibitively expensive.

The general classes of test equipment included are electrostatic detectors, surface resistivity meters, and static decay meters. Electrostatic detectors include electrostatic voltmeters, fieldmeters, and monitors.

Electrostatic monitors are used to detect a harmful electrostatic field and then sound an alarm or activate countermeasures such as air ionizers.

Electrostatic voltmeters and fieldmeters are used to measure the magnitude of electrostatic charge existing on materials, objects, or people. In addition, they can be used to measure the approximate magnitude of electrostatic charge generated by personnel movements and the triboelectric charge generated by rubbing two substances together.

These meters cannot detect rapid transients, that is, they cannot respond to pulses with fast rise times and short pulse widths. To measure such pulses it's necessary to use a high-speed storage oscilloscope, a discussion of which is beyond the scope of this text.

Three basic types of electrostatic detectors are used: electrometer, mechanical modulator, and nuclear.

In the electrometer a charge is induced on an electrode by the electric field; this charge, which is proportional to the field, is measured by a dc amplifier. The electrometer is adequate for quick measurements, but may have some drift, depending on the sophistication of the electronics.

In modulator type detectors a mechanical chopper such as a rotating propeller or slotted disc is placed between the electrostatic field and a capacitive sensor plate. The blades are kept at ground (zero) potential through contact with the case and person. As the propeller rotates, the blades and open segments alternately pass in front of the sensor (see Fig. 3-1). This action interrupts the field and thereby produces an electrical output on the sensor plate. The amplitude of this signal represents the intensity of the field; the signal's phase indicates the polarity of the field.[1]

In a typical nuclear electrostatic detector, the field induces into a grid, located in the tip of the meter, a charge that is proportional to the charge being measured. An ionizing source such as tritium foil causes the air space between the grid and a high impedance amplifier to become electrically conductive. As the air is ionized, it allows a small current to flow from the grid to the amplifier. This current is translated by the amplifier into the magnitude and polarity of the voltage being measured. The spacing between the object and the tip of the instrument determines the full scale sensitivity.[2]

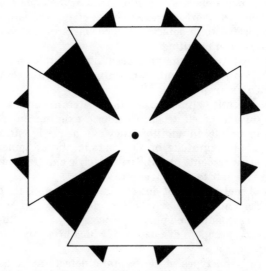

Fig. 3-1. Mechanical chopper for electrostatic meters (courtesy of R. E. Bolasny[1]).

Both electrostatic voltmeters and electrostatic fieldmeters are used for electrostatic measurements. While they have some basic similarities, they have some differences that should be noted.

Electrostatic voltmeters provide noncontacting measurement of the electrostatic surface potential (surface voltage) on insulators, semiconductors, or conductors. Their measurement accuracy is independent of probe-to-surface separation. Such meters can have high accuracy (0.1% or better) and the ability to resolve small spots, approximately 0.10 inch diameter or less.[3]

Electrostatic fieldmeters measure the electrostatic field (volts/cm) at the ground probe. For most applications, this is the field set up between the essentially grounded bottom plate of the probe and a charge surface some distance away. It is not the free space field that existed before the probe was inserted (except in certain circumstances where the probe views the field through or at the ground plane.) The field is proportional to the probe-to-surface separation.[3]

In contrast to electrostatic voltmeters, electrostatic fieldmeters have moderate accuracy, typically 2% to 5% plus, for measurement of surface voltage errors in measured probe-to-surface spacing. They have unlimited range for surface voltage measurements; since the range of the instrument is a function of the probe-to-surface separation, it is possible to measure hundreds of kilovolts by simply positioning the probe at an adequate distance from the surface under test.[3]

Electrostatic fieldmeters are used for measuring and monitoring of static charge accumulation on insulators (for example, in the manufacture of plastics, photographic film, paper, etc.) and for surface potential measurements where moderate accuracy and large area resolution are adequate.[3]

In selecting an electrostatic detector, the following characteristics should be considered:

1. Sensitivity in terms of minimum voltage level that can be accurately measured
2. Response time
3. Range of voltages that can be measured
4. Accuracy in various ranges
5. Radioactive or electrically operated
6. Portability
7. Ruggedness
8. Simplicity of operation and readability
9. Accessories such as remote probes and strip-chart recorder output.[4]

The instruments described herein are representative of the units available; inclusion here does not necessarily imply en-

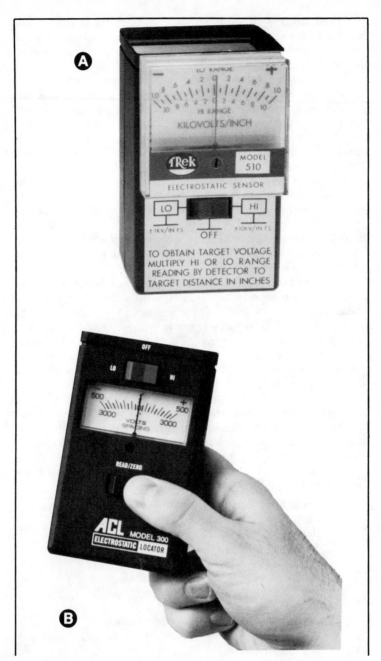

Fig. 3-2. Hand-held electrostatic meters ((A) courtesy of Trek, (B) ACL Incorporated, (C) Scientific Enterprises, and (D) Electro-Tech Systems).

dorsement. Lack of mention of a particular manufacturer or model does not mean that the unit is not as good, if not better, than the ones mentioned here.

3.1 PORTABLE ELECTROSTATIC METERS

For casual monitoring and for certifying ESD protected areas where Class 2 or Class 3 items (see Table 1-1) are handled, small portable electrostatic field meters may be used. (For certifying areas where Class 1 items are handled, more accurate laboratory-type detectors, described in section 3.2, may be required.)[4]

Portable electrostatic meters are available in two basic types: hand-held (Fig. 3-2) and suitcase-type bench meters. As most of the portable meters are the hand-held type, we will save discussion of the bench-type portable meters for section 3.2.

Hand-held units may be either pocket-type or pistol grip type. They are called by various names: static meter, electrostatic locator, static locator, field scanner, electrostatic detector, electrostatic meter, and static electricity detector/monitor.

In comparison to the bench-type meters, these meters provide coarse measurements, but they are nevertheless quite useful, in field use or in the plant, in determining the presence, polarity, and rough magnitude of electrostatic charges on various surfaces. Unlike conventional voltmeters, these meters are the noncontacting type. (Obviously, if they made contact they could discharge the surface whose potential was being measured.)

As hand-held meters, they are almost by necessity battery operated. Some have rechargeable type batteries. While considered portable meters, some of these units are designed for mounting on a camera tripod to allow for continuous readings.

Most of the units have analog displays. Scientific Enterprises Model 2B Field Scanner has a digital display. All of the units are nonradioactive except for the 3M Model 703 Static Meter which has a tritium foil as an ionizing source.

The ranges on these meters vary from a low of 1 to 50 volts to a high of 10,000 to 200,000 volts. Generally two or three scales or ranges are provided. Accuracy ranges from ± 5% to ± 10%.

The proper distance between the detector and the target varies with each meter, from 1 to 24 inches. It also may vary with different scales on the same meter. Failure to keep the specified distance can result in erroneous readings. For this reason, fixtures are sometimes used to maintain a uniform distance between the meter and the target.

With these meters it's essential to reference the meter to zero field before being used. Once the meter is in use, the operator must try to maintain a zero potential through the use of a wrist grounding strap (to be discussed in section 6.1). Otherwise a simple motion such as the operator scuffing his or her shoes may throw off the reading.

When measuring a surface, the surface should be in free air, away from a ground plane. Any ground plane next to (in back of), a surface will terminate the field and give erroneous readings on the field meter. That is, a triboelectrically charged surface will give different readings in free air and next to a ground plane, although the charge has remained the same.

3.2 PRECISION ELECTROSTATIC VOLTMETERS AND FIELDMETERS

Because of their accuracy, size, and expense, precision electrostatic voltmeters are considered laboratory type instruments. A voltmeter and a fieldmeter are described in the following paragraphs.

3.2.1 Electrostatic Voltmeters

The Monroe Electronics Model 244 electrostatic voltmeter[5] (see Figs. 3-3 and 3-4) permits accurate measurements of electrostatic or other high impedance sources without physical contact. It utilizes an electrostatic chopper for low drift and

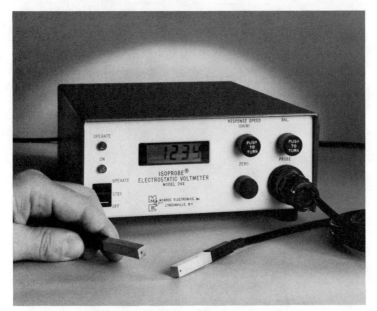

Fig. 3-3. Electrostatic voltmeter (courtesy of Monroe Electronics).

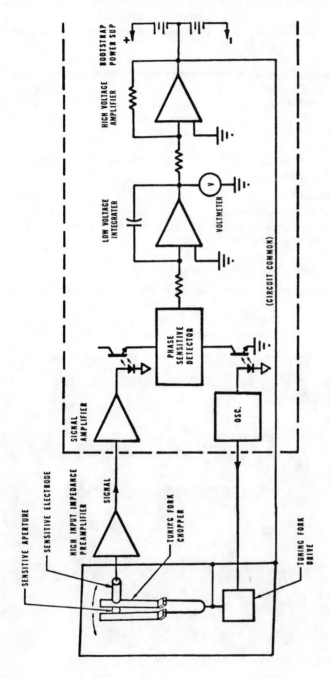

Fig. 3-4. Electrostatic voltmeter simplified block diagram (courtesy of Monroe Electronics).

60

negative feedback for accuracy. It has a range of 0 to ± 3000 volts dc.

Surface resolution is dependent upon probe aperture size and probe-to-surface spacing. Probe-to-surface spacing should be maintained as close as physically reasonable for best performance. Typical spacing range is from 0.005 inch for unknown voltages below 500 volts to over 0.125 inch for unknown voltages up to 5000 volts. As probe-to-surface spacing increases, instrument performance will suffer decreased accuracy, decreased speed of response, decreased surface resolution, increased noise, and increased drift.

When operating at probe-to-surface spacing of less than 0.125 inch, probe-to-surface arc-over may be a problem as air is subject to dielectric breakdown when the probe-to-surface spacing is low and the surface voltage is high. A destructive arc-over can occur, damaging the surface under test and/or the sensitive circuitry of the probe.

Principle of Operation (See Fig. 3-4). The electrostatic electrode "looks" at the surface under measurement through a small hole at the base of the probe assembly. The chopped ac signal induced on this electrode is proportional to the differential voltage between the surface under measurement and the probe assembly. Its phase is dictated by the dc polarity.

The reference voltage and this mechanically modulated signal, conditioned by the high input impedance preamplifier and signal amplifier, are fed to a phase-sensitive detector whose output dc amplitude and polarity are dictated by the amplitude and phase of the electrostatically induced signal relative to the reference signal. The output of the phase sensitive detector feeds a dc integrating amplifier. Its output polarity is inverted to that of the unknown. The output of this amplifier is used to drive a high voltage amplifier which in turn drives the probe to the same potential as that of the surface under measurement.

The probe is driven to a dc voltage typically within 0.1% of the potential of the unknown for a 0.040 inch probe-to-surface spacing. By simple metering the output of the high voltage amplifier, one has an accurate indication of the unknown potential.

Note that the Model 244 voltmeter is a noncontacting voltage follower. That is, the potential of its probe will attempt to follow the potential of any object within its field of view when the instrument is operating. Thus, if the meter indicates 1500 volts, for example, its probe will be operating at 1500 volts off ground.

In this case, if the probe is accidentally touched, it will apply *very briefly* 1500 volts to the unwary participant. However, the short circuit current that can be delivered by the meter is limited to approximately 500 microamperes which is below the threshold of feeling for most people and therefore is clearly no hazard. Typically a spark will jump which will be disturbing but not hazardous electrically. (But you could recoil from the surprise and hurt yourself.) The spark is caused by the discharge of the capacitance of the system through the body. This stored energy is typically only a few millijoules while the lethal level of stored energy has been estimated as 27 joules.

Probe Maintenance. It is impossible for the sensitive aperture in the probe to be covered by a window of any sort which will permit the probe to operate normally. Therefore, so long as this aperture is open, foreign material may enter the volume associated with the vibrating vane and generate undesired noise or dc offset, thereby impairing the accuracy and utility of the instrument.

To maintain low noise and low offset, it is important that the probe be operated in as clean an environment as practicable, purged with filtered air or an inert gas such as clean, dry nitrogen, and periodically disassembled and cleaned. Where practicable the probe should be installed with the sensitive aperature downward and when not in use, wrapped with aluminum foil.

Gradient adapters are available as accessories to permit use of the Model 244 as an electrostatic fieldmeter with direct readout in volts per millimeter of separation between the surface under test and the grounded plate of the adapter. The adapter is a gold-plated metal plate one inch square with an aperture through which the probe "looks", a 10-foot flexible grounding wire with clip, and a small insulated box designed to slip over the end of the probe.

3.2.2 Electrostatic Fieldmeters

The Monroe Model 171 electrostatic fieldmeter[5] measures electrostatic field strength or intensity (potential gradient) in kilovolts per centimeter up to $\pm 10\,kV/cm$). It may also be used to determine surface voltage by using the probe-to-surface separation as a calibration factor. See Figs. 3-5 and 3-6.

The instrument is expandable from 2 to 16 channels. Each channel has two alarms brought out separately, one to indicate a plus field beyond the set limit and the other a minus field beyond its limit. These alarm lines are intended for activation of alarm relays. A meter is provided only for convenience of observation as the Model 171 is designed for feeding data to a central data collecting/monitoring point via the analog outputs.

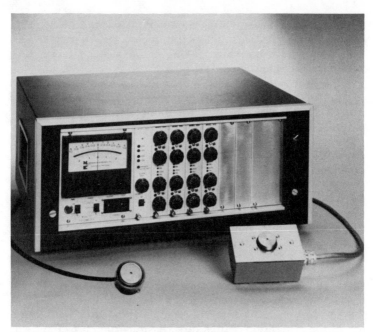

Fig. 3-5. Electrostatic fieldmeter (courtesy of Monroe Electronics).

A major application of the Model 171 is the monitoring of electrostatic charge accumulation at a number of points simultaneously with one instrument and up to 16 probes. As charge increases on the surface of a material the electrostatic field in the

Fig. 3-6. Electrostatic fieldmeter (courtesy of Trek).

vicinity also increases proportionately. Thus the Model 171 provides an output signal directly proportional to the surface charge accumulation while making no physical contact to the material being monitored.

The probes are intended for use at any location requiring monitoring of charge build up. Probes can be separated from the main frame by up to 1000 feet. The probe should be in proximity to the surface under measurement.

Principle of Operation (See Fig. 3-7). A vibrating sense electrode (capacitor) "looks" at the surface under measurement through a sized hole in the gradient cap. An ac signal is induced on the vibrating electrode, which is capacitively coupled to the surface under measurement; thus, not requiring contact. This signal's amplitude and phase are dictated by the size and polarity of the dc voltage on the surface.

This ac signal, conditioned by the preamplifier, filter and signal amplifier, is fed into a phase sensitive demodulator. The signal from the demodulator then feeds an integrator. A voltage is then fed back to the gradient cap to null the field. When the field becomes null, the integrator output stabilizes. The voltage out of the integrator is thus directly proportional to the field intensity at the probe. This makes it possible to couple to a meter for direct readout.

3.3 ELECTROSTATIC MONITORS

Electrostatic monitors are designed to sound an alarm when the electrostatic potential reaches a predetermined danger level, thereby alerting personnel before harm occurs. While these monitors are available in single self-contained units, they are

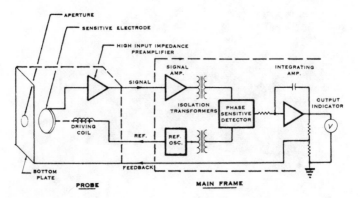

Fig. 3-7. Electrostatic fieldmeter simplified block diagram (courtesy of Monroe Electronics).

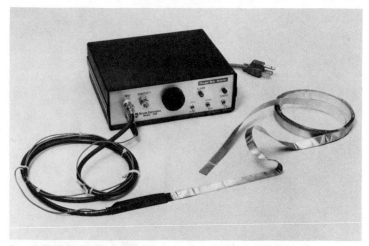

Fig. 3-8. Electrostatic monitor (courtesy of Monroe Electronics).

frequently multistation, that is, they may have several sensors connected to one central control. The sensors may be as far as 1,000 feet from the main console. In addition to sounding an alarm, the console may activate air ionizers and provide a strip-chart record of the event.

Another type of electrostatic monitor (Fig. 3-8) detects static events rather than a slow charge accumulation. It gives a pulse output each time a charged object crosses its tape sensor. The primary use of such monitors is to evaluate the effectiveness of static control techniques.

3.4 SURFACE RESISTIVITY PROBE

The Electro-Tech Systems Model 802 surface resistivity probe[6] (see Fig. 3-9) is used to measure the surface resistivity of relatively smooth, flat materials believed to be anti-static or static-dissipative. These materials include workbench tops, conductive floors, tote boxes, antistatic film, etc.

Materials to be measured must have a total area within which the 5-inch diameter probe can be positioned. Critical dimensions of the probe provide for a multiplication factor of ×10 applied to the measured resistance reading. The dimensions have been derived from applicable formulas set forth in ANSI/ASTM D 257-78.

The probe uses the guard-ring principle of measurement which excludes all areas of the material under test except the portion located between two contact rings. The contact surfaces consists of two concentric circles composed of monel braid woven over a neoprene sponge core having a square cross section.

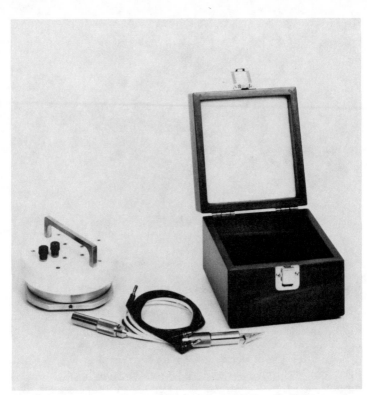

Fig. 3-9. Surface resistivity probe (courtesy of Electro-Tech Systems).

The braid is compressed against the surface under test by the weight of the probe, compensating for any slight unevenness and providing a firm electrical contact around the entire circumference of each ring. A ground plane is provided for measuring the surface resistivity of films and thin sheets.

The probe can be used with any standard high resistance ohmmeter which has a charging voltage of 500 volts. The meter should be capable of reading up to at least 10^{12} ohms. With the $\times 10$ multiplication factor of the probe, surface resistivities up to 10^{13} ohms per square can be measured.

It's important to check surface resistivity at lower voltages as well. With many materials the contact resistance is such that it provides a high resistance at low voltage and much lower readings at high voltage. For example, when a charged tote box is placed on a table, it will discharge through the table top until its voltage has decayed to the point where it can no longer break down the contact film that gives rise to the high contact resistance. This tends to limit the decay level that can be achieved.[7]

3.5 STATIC DECAY METER

A static decay test measures the ability of a material, when grounded, to dissipate a known charge that has been induced on the surface of the material. It is a method to evaluate antistatic-treated materials as described in Federal Test Method Standard 101B—Method 4046, "Antistatic Properties of Materials."

The Electro-Tech Systems Model 406C static decay meter[6] (see Fig. 3-10) is a completely integrated system for measuring the electrostatic properties of materials in accordance with Method 4046. It can be used to analyze the effectiveness of antistatic additives and sprays.

The meter measures the time required for a charged test sample to discharge to a known, predetermined cutoff level. Three manually selected cutoff thresholds at 50%, 10%, and 0% of full charge are provided. A regulated high voltage power supply contained within the unit, in conjunction with self-contained precision charging voltage and sample charge meters enables the sample charge to be adjusted to any level from zero to $\pm 5\,kV$.

The sample is contained in a special Faraday cage, which includes a patented electrostatic voltmeter which enables the system to make an electrostatic (noncontact) measurement of the charge on the sample. The cage shields the test sample from any outside electrostatic interference.

A calibration module is supplied with the Model 406C to enable the user to verify the correct operation of the instrument. This module contains electronic circuitry that simulates the static decay characteristics of a test sample.

3.6 HUMIDITY TEST CHAMBER

The Electro-Tech Systems Model 506 humidity test chamber[6] provides a controlled humidity for performing those tests

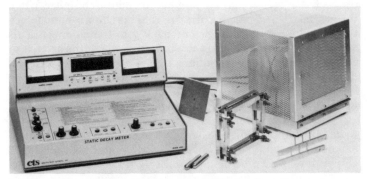

Fig. 3-10. Static decay meter (courtesy of Electro-Tech Systems).

requiring special humidity conditions, such as the static decay test for measuring the electrostatic properties of materials.

Materials that are inherently not static-free, such as plastics and nonwovens, can be treated with topical or internal antistats to make them static free. In the case of topical or internal treatment, the effectiveness of the antistat is a function of the relative humidity since one of the purposes of the antistat is to draw moisture from the air to provide a conductive surface film.

Depending on the end application of a product, a material may only have to be static free at moderate humidities such as 50%, or have to be static-free down to relative humidities of less than 15%.

Military specification MIL-B-81705B specifies the antistatic requirements for materials used to package electronic components. It calls for such materials to be tested at a relative humidity of 15% or less at a temperature of 73 degrees F $\pm$ 3 degrees after stabilization at these conditions for at least 24 hours.

The ability to obtain the correct humidity and temperature conditions within the test chamber is a function of the surrounding conditions. The most difficult condition to achieve is to bring the humidity in the chamber down to 15% when the outside humidity is quite high (over 50%). To accomplish this, the ETS Humidity Test Chamber is designed to dry the moist air within the chamber with a circulating pump and desiccator.

The chamber is an airtight acrylic glove box. Access to the box is through a square door-opening located on the side and held closed by four latches. Samples placed inside the box are handled by the operator placing his hands in the neoprene gloves located in the front.

The chamber has two desiccators for drying. One is a passive drying column which is placed in the corner of the chamber. The other desiccator is mounted externally and is connected in series with a small pump and the chamber. When the pump is turned on, air is drawn from the chamber and forced through the external desiccator back into the chamber. This circulating of the air inside the chamber through the desiccator can produce very low humidities in a relatively short period of time. Once the chamber has been brought down to the desired humidity level, the passive desiccator will maintain the level without having to continuously operate the pump.

A combination temperature/humidity indicator is provided to monitor the conditions within the chamber. An optional extender is available so that extra long materials such as IC shipping tubes can be nondestructively tested in the chamber.

4

Protective Packaging

W HENEVER ESD SENSITIVE DEVICES ARE NOT BEING WORKED ON, THAT is, when they are transported or stored, they must be kept in electrostatic protective packaging. Note that this transportation can be across a room as well as a transcontinental trip. ESD protective packaging includes bags, tote boxes, DIP tubes, conductive shunts (shorting plugs), conductive foam, and cushioned packaging material.

It is difficult to select the best packaging material because of the many products on the market and the questionable claims made by rival manufacturers.[1] This chapter gives the basic principles that should be considered in selecting such packaging.

Because of the sometimes contradictory claims by manufacturers, any major purchase of ESD protective materials should be made only after a thorough review of the market, a formal product qualification program, and periodic lot sample testing.[2]

Static protective containers must provide protection (1) against triboelectric generation, (2) from electrostatic fields, and (3) against direct discharge from contact with charged people or a charged object. Because it is difficult to find one material with comprehensive protection, it is often necessary to use a combination of different protective materials to obtain proper protection.

Static protective materials are generally identified as either conductive (10^5 ohms/square or less), static dissipative (10^5 to 10^9 ohms/sq), or antistatic (10^9 to 10^{14} ohms/sq). As to the level of protection provided by each type, there is considerable controversy between manufacturers of antistatic materials and manufacturers of conductive materials. Both types have proved useful. Because of this, cost may be the key consideration in selecting such materials. In general the cost of protective materials increases as the conductivity goes up.[3]

Table 4-1 compares these three classifications of ESD protective materials. Note that the information is relative from one material type to another. It is important to note too that there are no sharp dividing lines between the three types of materials. For example, the properties of a conductive material at the higher end of its resistivity range could be equivalent in properties to a static dissipative material at the lower end of its resistivity range.

As triboelectric generation is a friction process, one of the prime ways to reduce it is to increase the material's lubricity, which is a measure of surface smoothness and the lubricating action of moistness. The higher the lubricity of the surfaces being rubbed, the lower the friction and therefore the lower the generated charges.

Antistatic materials are impregnated with antistats which constantly migrate to the surface forming a sweat layer that increases the material's lubricity. These hygroscopic materials attract moisture from the surrounding air. If the relative humidity is low, their effectiveness will decrease. Periodic tests may be necessary to ensure that it has not deteriorated. One study showed that most antistatic plastics will lose effectiveness from contact with paper products or exposure to the air.[4]

Conductive materials are typically impregnated or loaded with carbon to produce volume-conductive materials. Such materials are widely used as electrostatic shields (Faraday cages) to minimize induced charge effects of electrostatic fields. As the shielding ability of a material is directly related to its conductivity,[5] antistatic materials cannot be used as electrostatic shields.

Electrostatic shields also provide protection against the destructive electromagnetic pulse which might otherwise be induced in an enclosure from an ESD high voltage spark.

For an electrostatic shield to be effective it must form a complete conductive cage around the item to be protected. An open tote box, that is, one without a top, does not form a complete cage and therefore cannot completely shield against an electrostatic field.[6]

In addition to providing protection against electrostatic fields, many conductive and static dissipative materials also provide protection from triboelectric generation. However, some

Table 4-1. Comparison of ESD Protective Types (DOD-HDBK 263).

Conductive	Static Dissipative	Anti-Static
1. Could present a personnel safety hazard when contacting high voltages and hard grounds.	1. Presents the same hazards as listed under "Conductive" 1 and 2 except to a lesser degree. Hazards depend upon the magnitude of the voltages and the types of parts and circuits tested.	1. The effectiveness of hygroscopic antistatic materials are reduced in low relative humidities since their anti-static properties are dependent upon absorbing moisture from the air.
2. Could damage electrical circuitry of parts or assemblies during testing if electrical connections contact conductive surfaces.	2. See item (6) under "Conductive".	2. The accumulation of dirt, oils and silicone have an adverse effect on the anti-static properties of hygroscopic antistats. Cleaning with solvents such as alcohols, ketones and other hydrocarbon based solvents can remove the antistats. May require periodic treatment with a topical antistat.
3. Steels (except corrosion resistant) are prone to corrosion. Protective coatings such as paint will destroy the surface conductive properties and could be static generative.	3. See item (7) under "Conductive".	3. Antistats used in some hygroscopic anti-static materials can track onto items and act as a foreign substance which could react with other materials adversely. This has been shown to be a problem with the lubricant in miniature bearings.
4. Aluminum will form aluminum oxide on its surface reducing conductivity and increasing its ability to generate static.		4. See item (6) under "Conductive".
5. Hard surfaces such as metal provide little protection from physical shock to items dropped thereon		5. Hygroscopic anti-static materials generally provide protection against triboelectric generation. The triboelectric generation characteristics of other anti-static materials depends upon the material used.
6. Materials should be reviewed for flammability, corrosivity, toxicity, bacterial growth, crumbling, powdering, shedding, flaking, brittleness, outgassing, long term chemical reaction with parts.		
7. Protection against triboelectric generation depends upon the material used.		

metals will create significant charges from triboelectric generation as is indicated in the triboelectric series. For example, aluminum can generate substantial electrostatic charges when rubbed with a common plastic. Although a conductive material distributes a charge over its surface, the other substance with which it is rubbed or from which it is separated, especially if it is insulative, can become highly charged.[7]

Whether conductive, static dissipative, or antistatic, ESD protective materials can be formed into numerous shapes. Metals, for example, can be cast, stamped, or piece welded into most any shape while most conductive, static dissipative and antistatic plastics can be molded into formed shapes. Fiberboard, melamine laminates, and similar materials in laminated or homogenous form can be constructed into boxes and various other shapes. These shapes include the following:

1. Sheets and plates in various sizes and thicknesses for use as workbench tops, floors, floor mats, wraps, and coverings.
2. Formed parts trays, vials, carriers, boxes, bottles, and other custom shapes.
3. Rigid shorting bars and clips for electrical shorting of ESDS part leads and higher assembly connectors.
4. Foam used for shorting part leads, assembly connectors, or as a cushioning for packaging.
5. Bubble pack material or open cell plastic foam—used for a cushioning for packaging such as MIL-P-81997 and PPP-C-1842 Type III Style A.
6. Flexible materials in the form of bags such as MIL-B-82467, trash can liners, seat covers, and personnel apparel including smocks, gauntlets, and finger cots.
7. Personnel ground straps—insulated wire or flexible straps of the conductive plastic used in the form of personnel ground strap cables.
8. Heel grounders—flexible formed strips of the conductive plastics placed inside the shoe and connected to the outside shoe heel used to ground personnel to the bottom of the shoe heel.
9. Conductive shoes—conductive rubber or plastic inner and outer soles and heels.
10. Conductive, static dissipative, and antistatic carpeting and flooring, including carbon impregnated vinyl tiles and terrazzo floors.[7]

As the controversy between proponents of conductive materials and antistatic materials continues, it appears that both groups are moving closer to the middle ground of static dissipative materials.[8] In the meanwhile the following summary of these materials should be kept in mind.

In general, conductive materials will:

1. Be opaque (probably black).
2. Cost more than antistatic.
3. Require no maintenance.
4. Stay within specification until it wears out.
5. Perform independently of moisture or relative humidity.
6. Accept printing.

Antistatic materials will:

1. Be opaque or translucent.
2. Cost less than conductive materials.
3. Require periodic tests to determine their functionability.
4. Depend upon relative humidity to perform.
5. Will not accept printing.[8]

In selecting ESD protective materials, one should note an additional precaution: there may be significant quality variations in antistatic packaging materials. One researcher has found that static decay times vary from lot to lot from individual manufacturers and vary even more dramatically between manufacturers.[9]

Regardless of the type of material, all protective packaging must have an obvious label that states electrostatic sensitive devices are contained therein. Sensitive devices must not be removed from their containers except at static-free work stations, and then only at the actual time they are in work. Before removing an item from its protective container the operator should (1) place the container on a conductive grounded bench top, (2) make sure the wrist strap fits snugly around the wrist and is properly plugged into the ground receptacle, and (3) touch hands to the bench top.[10]

With these precautions in mind, let's consider each type of protective container.

4.1 PROTECTIVE BAGS

Three types of ESD protective bags are commonly available: carbon-impregnated polyolefin film which is a black, opaque bulk conductive plastic; transparent pink antistat impregnated plastic, commonly known as "pink poly", and metallic films laminated with other materials. See Fig. 4-1.

The advantage of a transparent bag in comparison with an opaque bag is that the contents of the bag can be easily identified without removal from the bag. Such identification is frequently needed in incoming inspection to verify the contents of the bag. Each time an ESDS device is removed and reinserted in a bag adds to the probability of it being accidentally damaged.

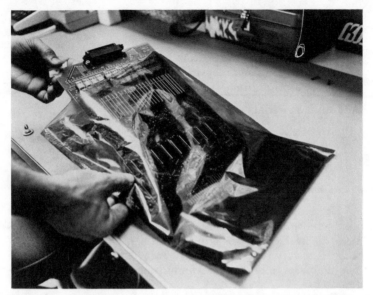

Fig. 4-1. ESD protective bag (courtesy of Static Control Systems/3M).

Bulk conductive plastic bags are generally restricted from use in clean rooms because they tend to smudge.[11] Another problem with these bags is that they are dangerous near live electrical circuits because of their conductivity.[12]

However, some bulk conductive plastic bags provide some protection against external electrostatic fields.[11] Because they are conductive, they act as Faraday cages as shown in Fig. 4-2.

Pink poly bags provide protection against triboelectric charging but no protection against external electrostatic fields.

The laminated or sandwich-construction bags are available as opaque or transparent bags. The opaque bags have a layer of aluminum foil that protects objects from electromagnetic interference and also provides an effective moisture vapor barrier. At least one type has antistatic surfaces, both interior and exterior.

Figure 4-3 shows the construction of a typical transparent bag that has a metallized electrical barrier. The metal used here is not stated but in other products one of the layers is vapor coated with nickel. In either case, the interior layer is antistatic to prevent triboelectric charging. These transparent bags provide protection from electrostatic fields but not from electromagnetic interference.

4.2 DIP TUBES

Dual in-line package (DIP) type integrated circuits are usually shipped in tubes commonly referred to as DIP sticks (see

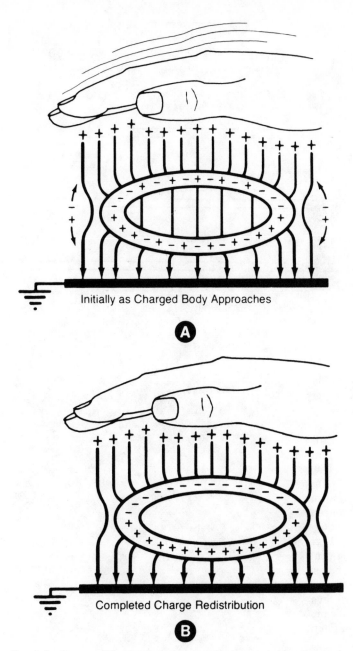

Initially as Charged Body Approaches

A

Completed Charge Redistribution

B

Fig. 4-2. Charge redistribution on a conductive container (courtesy of Donald M. Yenni, Jr., copyright American Society for Quality Control, Inc. Reprinted by permission).

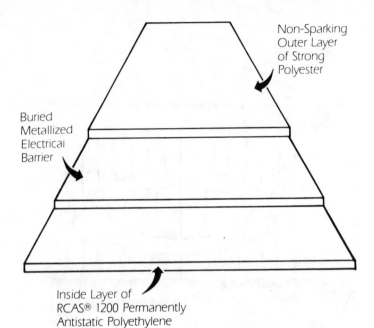

Non-Sparking
Outer Layer
of Strong
Polyester

Buried
Metallized
Electrical
Barrier

Inside Layer of
RCAS® 1200 Permanently
Antistatic Polyethylene

Fig. 4-3. Laminated ESD protective bag construction (courtesy of Richmond Division of Dixico Incorporated).

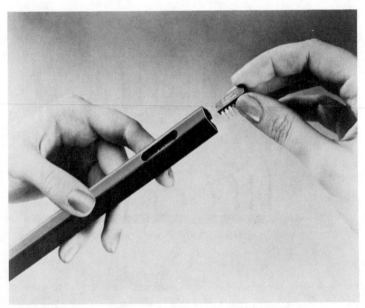

Fig. 4-4. DIP stick (courtesy of Static Control Systems/3M).

76

Fig. 4-4). During shipment the devices move slightly in the sticks and thereby acquire a charge. When the stick is emptied, each device is likely to be grounded. At that time the rapid discharge may damage or destroy the device.[13]

As the static charge generated on a device when sliding in a shipping tube is due to different materials rubbing against one another, in order to minimize charge build-up the two items should be made from the same substance or from materials that lie close to each other in the triboelectric series.[13]

Tests at Bell Laboratories showed that aluminum sticks without an anodic coating provided the least charge on all

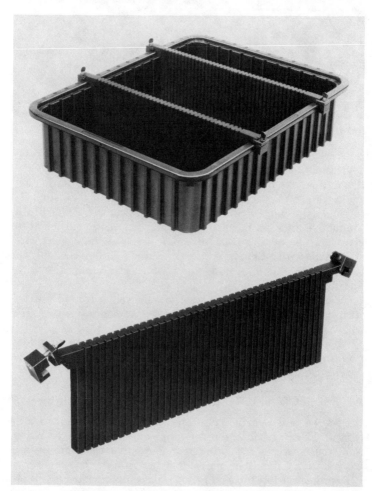

Fig. 4-5. ESD protective tote box (courtesy of Static Control Systems/3M).

Fig. 4-6. ESD protective component tray (courtesy of LNP Corp.).

devices. Other sticks, in descending order of effectiveness were Benstat® sticks, new antistatic coated PVC sticks, carbon loaded sticks, clear PVC sticks, and used antistatic coated PVC sticks.[13]

Only the more ESD sensitive devices must be caged in a conductive stick. In most tubes the clearances are sufficient to give some protection from magnetic fields. A new composite stick made with a Benstat® inner liner and a carbon loaded PVC outer sheath should provide good triboelectric properties with Faraday caging for sensitive devices.[13]

For device protection, a person should wear a grounded wrist strap when handling sticks loaded with devices, particularly with conductive sticks.[13]

Some DIP sticks have window viewing slots to allow counting and reading of device type without removing the devices from the stick.

4.3 PROTECTIVE TOTE BOXES AND STORAGE BINS

As we noted earlier, protective bags are essential in protecting ESDS devices during shipment and storage. When numerous devices are being transported between static-safe work stations, it may be more convenient to carry the devices in ESD protective tote boxes (Fig. 4-5) and component trays (Fig. 4-

6). In storage and handling, ESD protective storage bins and cabinets (Fig. 4-7) prove quite useful.

Like ESD protective bags, ESD protective tote boxes and storage bins are made of either conductive or antistatic materials. By necessity, some of these containers must have open tops, which leaves the devices vulnerable to external static fields. Some conductive tote boxes have hinged lids giving a complete Faraday cage protection for the devices.

As with protective bags, the greater the conductivity the greater will be the protection against external fields. If, however, the conductivity is too great the static decay time will be too short and a spark discharge will occur.

Fig. 4-7. ESD protective component storage cabinet (courtesy of Stanley-Vidmar).

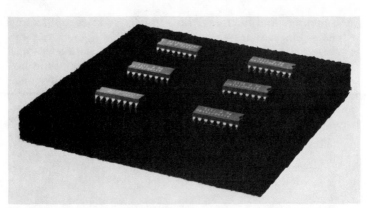

Fig. 4-8. Conductive foam (courtesy of Static Control Systems/3M).

4.4 PROTECTIVE FOAM

Ordinary plastic foams readily produce static charges and therefore should be kept away from ESDS devices. However, antistatic and conductive foams are widely used for protecting static sensitive devices during shipment.

Foams prevent bending or misalignment of device leads, absorb vibrations and physical shocks, and short device leads to provide equipotential bonding, as shown in Fig. 4-8. Low density foam provides optimum cushioning while high density foam

Fig. 4-9. Conductive board shunt (courtesy of Static Control Systems/3M).

gives maximum shorting of IC leads.

It's important to note that foams do not give complete protection to ESDS devices. For example, if a statically charged person touches a floating metal lid of a DIP, an arc can form between the lid's inside surface and the chip's top surface.[14]

4.5 CONDUCTIVE SHUNTS

Conductive shunts (see Fig. 4-9) are designed to slide over printed circuit board edge connectors. As the shunt connects all the leads together, it provides equipotential bonding, which helps protect ESDS parts while the board is being handled or shipped.

5

Protective Environment

━━━━━━━━━━━━━━━━━━━━━━━━━━━━━━━━━━━━━━

AS WE HAVE SEEN IN THE PREVIOUS CHAPTER, PROTECTIVE PACKAGING is essential in protecting ESDS items. When an ESDS item must be removed from its packaging for assembly or processing, it is most important that the environment near the item be altered to protect the item. This can be accomplished through humidity control, special work surfaces and floors, ionized air, and proper tools.

5.1 HUMIDITY CONTROL

At high humidity levels, a thin layer of moisture forms on the surface of objects in the area. This surface film provides a conductive path for any charge that may exist or is generated on the surface, thereby dissipating any electrostatic charge. In addition, moisture in the air tends to neutralize surface charges. It also adds lubricity to prevent frictional charging. In general, the lower the humidity, the higher will be the electrostatic potential, as shown in Table 5-1.

For this reason, adding moisture to the air is one way of combatting electrostatic charges. The more moisture that is added, the more likely it will be that electrostatic charges will dissipate. But as shown in Table 5-1, even at the highest levels, significant electrostatic voltages can not be eliminated. Also, at very high humidity levels, workers can be very uncomfortable, parts can rust, electric leakage paths can form, and printed wiring boards can delaminate during soldering.

The optimum relative humidity level in ESD protective areas, considering the above factors, has been specified as either 40

Table 5-1. Effect of Humidity on Electrostatic Voltages (DOD-HDBK 263).

Means of Static Generation	Electrostatic Voltages	
	10 to 20 Percent Relative Humidity	65 to 90 Percent Relative Humidity
Walking across carpet	35,000	1,500
Walking over vinyl floor	12,000	250
Worker at bench	6,000	100
Vinyl envelopes for work instructions	7,000	600
Common poly bag picked up from bench	20,000	1,200
Work chair padded with polyurethane foam	18,000	1,500

percent to 50 percent[1] or 40 percent to 60 percent[2] depending on which authority you choose. For some climates at certain times of the year it can be quite difficult and expensive to obtain a 60 percent level. In such cases, air ionizers (see section 5.4) offer an effective substitute or supplement for dissipating electrostatic charges.

5.2 WORK SURFACES[2]

Work surfaces such as the tops of workbenches that contact ESDS items and personnel should be ESD protective over the area where ESDS items may be placed. These protective work surfaces are an important defense against ESD damage. The purpose of such surfaces is to drain static charges from any conductor placed on them; nonconductors, of course, will not be neutralized by such surfaces. The discharge should take place rapidly so as to prevent damage to ESDS items but slowly enough to prevent an arc discharge which could also cause damage.

Three categories of materials are used for these surfaces: conductive, static dissipative, and antistatic. These categories are based on DOD Handbook 263's definition of resistivity: conductive, less than 10^5 ohms per square; static dissipative, between 10^5 ohms and 10^9 ohms per square; and antistatic, between 10^9 and 10^{14} ohms per square.

The characteristics of each category are given in Table 5-2. Considerable controversy exists over which material is the best. The Reliability Analysis Center has concluded that *static dissipative* materials are ideal because antistatic table tops take too long to dissipate a static charge and conductive table tops dissipate static charges too quickly.[3]

Table 5-2. Types of ESD Protective Bench Tops (DOD-HDBK 263).

Conductive	Static Dissipative	Antistatic
1. Dissipates charges rapidly throughout the material and to ground, and will not maintain a high static voltage.	1. Charge dissipation rate generally adequate for most ESDS parts.	1. Provides slow bleedoff of static charges. If ground straps are used by personnel working at the work bench high ESD voltages should be rapidly dissipated through the ground strap.
2. Could discharge an ESD in the form of a spark causing EMP.	2. Provides greater resistance for personnel protection from high voltages or hard grounding if the table top is contacted with test equipment ground.	2. Eliminates sparks from ESD.
3. Could cause a high current discharge through an ESDS part.	3. Reduces discharge currents through ESDS parts.	3. Limits discharge currents through ESDS parts to low levels.
4. Could present a safety hazard or short if a high voltage source contacted the bench top. Could hard ground the table top if test equipment with grounded chassis contacted the bench top surface.	4. Safety could require that series resistances be provided in connection to ground where high voltages can be contacted by personnel.	4. Generally provides adequate resistance for personnel safety.
5. Safety could require that series resistances be provided in connection to ground where high voltages can be contacted by personnel.		

Personnel wrist straps (see section 6.1) should be used with protective work benches to prevent personnel from discharging through an ESDS item to the workbench surface.

Workbench surfaces should be connected to a *soft* ground, that is, through a connection to ground that is in series with a specified resistor. This resistor should be located at or near the point of contact with the workbench top. It should be large enough to limit any leakage current to 5 milliamperes or less considering the highest voltage source within reach of grounded people and all parallel resistances to ground, such as wrist straps, table tops, and conductive floors. Typically this resistance is 1 megohm. Protection to less than 5 milliamperes may be advisable where reflex action could cause problems. Fig. 5-1 shows a typical ESD grounded workbench.

5.2.1 Grounding Considerations[2]

In the construction of ESD protected areas and at grounded workbenches, several steps are necessary to reduce the chance of

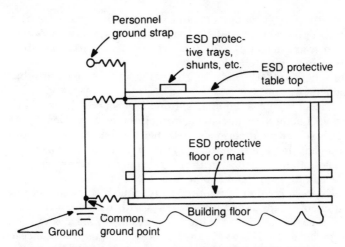

Fig. 5-1. Typical ESD grounded work bench (DOD-HDBK 263).

electrical shock to personnel. Before we consider these, however, let's look at the physiological effects of electrical shock.

The severity of electrical shock is determined by the magnitude and path of the current flowing through the body and the duration of time that the current flows. As shown in Table 5-3, even relatively small currents can be fatal if the path includes a vital part of the body, such as the heart or lungs. The voltage necessary to produce a fatal current depends upon the resistance of the body, contact conditions, and the path through the body. In addition to the danger of shock, there is the danger of electrical burns which are usually produced by heat from the arc that occurs when a body touches a high-voltage circuit. Electrical burns are also caused by current passing through the skin and tissues.

5.2.2 Grounding Electronic Test Equipment and Tools[2]

Because of shock hazards in ESD protected areas and on ESD grounded work benches, all external parts, surfaces, and shields

Current Values (Milliamperes)		Effects
Ac 60 Hz	Dc	
0-1	0-4	Perception
1-4	4-15	Surprise
4-21	15-80	Reflex action
21-40	80-160	Muscular inhibition
40-100	160-300	Respiratory block
Over 100	Over 300	Usually fatal

Table 5-3. Effects of Electrical Current on Humans (DOD-HDBK 263).

in electronic test equipment and power tools need to be at a common ground potential at all times during normal operation. Where a ground is part of the circuit, any external or interconnecting cable should carry a ground wire in the cable terminated at both ends in the same manner as the other conductors.

Except with coaxial cables, in no case should the shield be depended upon for a current-carrying ground connection. Plugs and convenience outlets for use with metal-case portable tools and equipment should automatically ground the metal frame or case of tools and equipment when the plug is mated with the receptacle; the grounding pin should make first, break last.

Extreme caution is required in placing such tools and test equipment on grounded workbenches with metal or other conductive coverings since the hard case ground of the tool and test equipment grounded cases can shunt the protective resistance in the workbench ground cable.

For further protection for personnel, ground fault interrupters (GFI) should be used with test equipment. The GFI senses leakage current from faulty test equipment and instantly interrupts the circuit when the current reaches a potentially hazardous level.

It is most important to avoid parallel paths to ground that could reduce the equivalent resistance of persons to ground to unsafe levels. Such parallel paths could result from the use of wrist straps, grounded table tops, and grounded floor mats.

5.2.3 Ground Potential of Grounded Workbenches

Additional safety and grounding considerations for ESD protected areas and grounded workbenches are as follows:

1. Cables and resistors should have ample current carrying capacity. Since the purpose of the workbench ground is to bleed off electrostatic charges, a half-watt resistor is usually sufficient.
2. Ground cable connections should be continuous and permanent.
3. The ground cable and connection material should be strong enough to minimize accidental ground disconnections.
4. Workbench tops, floor mats, ground straps, and other ESD protected area grounds used to discharge static electricity should be connected to earth, power system, or other hard ground as appropriate, through current-limiting resistances. The wrist strap should be connected to ground through the ground point of the work-

bench top (see Fig. 5-1.) Workbenches should not be connected in series with one another because the series resistances will add together, giving a higher ESD dissipation time. Also, an opening in one ground cable could affect the other workbench ground cables.

Safety considerations specify the *minimum* resistance to ground for protection of personnel; the *maximum* resistance to hard ground for personnel grounding is determined by the decay time for an electrostatic charge. This decay time is determined by the human capacitance and resistance and the resistance of other ground paths to hard ground. It should be short enough to dissipate charges at or below the rate at which they are normally generated.

5.2.4 Mechanical Considerations

In addition to electrical considerations, several mechanical considerations must be noted in choosing material for a workbench top. If the surface is too hard, parts can be damaged if they are dropped on it. But if it is too soft, its durability will suffer. The surface needs to be resistant to chemicals such as trichlorethylene that are frequently found in electronics assembly and test areas.

The purpose of the workbench may dictate which of these factors will be most important; a workbench for PCB repair may not be appropriate for chassis assembly.[4]

ESD protective surfaces are available as a permanent part of a workbench or as removable mats, as shown in Figs. 5-2 and 5-3.

5.3 ESD PROTECTIVE FLOORS

Just about any type of floor, whether it is painted or sealed concrete, finished wood, vinyl tile, or carpet, can be a prime contributor to ESD problems. Consequently, floors should be either covered with ESD protective flooring or floor mats or be treated with a topical antistat, if complete ESD protection is needed.

Like ESD protective workbench tops, protective flooring is available as either conductive, static dissipative, or antistatic materials. The characteristics of each type are given in Table 5-4. These materials are available in the form of floor mats, vinyl floor tiles, and terrazzo.

Conductive vinyl floor tiles have long been used in hospital surgery rooms to prevent explosions caused by ESD. This tile is designed to conduct charges to ground and to conduct from tile to tile.

Fig. 5-2. ESD protective table mat (courtesy of SIMCO).

The disadvantages of floor mats (Fig. 5-4) are that whenever a floor is cleaned they must be moved and they can be a tripping hazard if not carefully placed. Just as with ESD protective

Fig. 5-3. ESD protective table mat in place (courtesy of Static Control Systems/3M).

Fig. 5-4. ESD protective floor mat (courtesy of Static Control Systems/3M).

workbench tops, conductive floors or floor mats must have a current-limiting resistor in the ground lead.

When conductive floors or mats are used, personnel should wear either conductive shoes, shoe covers, or heel grounders as shoes of man-made materials may have inadequate conductivity. These items should be kept clean for if they become contaminated by dirt or chemicals they will lose their conductive interface with the floor. Because of safety, they should not be worn outside an ESD protected area.

Conductive work stools are necessary with protective flooring because a worker sitting at a workbench is likely to lift his feet from the floor to the work stool; if the stool were not grounded, the benefits of the protective flooring would be lost.

Conductive floor finishes may be mopped or painted onto floors to provide ESD protection. Because of foot and cart traffic, these finishes must be periodically reapplied. Whatever protection is used, protective floors must not be waxed as the wax buildup will reduce the conductivity.

Table 5-4. ESD Protective Floor Mats Characteristics (DOD-HDBK 263).

Conductive	Static Dissipative	Antistatic
1. Dissipates charges rapidly throughout the material and to ground, and will not maintain a high static voltage. 2. Safety could require that series resistances be provided in connection to ground where high voltages can be contacted by personnel.	1. Provides adequate conductivity for dissipation of charges. 2. Generally provides sufficient resistance for personnel safety. External series resistance to ground may not be required.	1. Provides slow bleed-off of high static charges. 2. Accumulations of dirt, contaminants and wear reduce anti-static properties. Requires frequent cleaning and treatment with a topical antistat.

5.4 AIR IONIZERS

In previous paragraphs we have noted that conductive items in the electronics work environment should be grounded through wrist straps and protective work surfaces. In some cases nonconductive items can be given a conductive path, for dissipating the electrostatic charge, by water vapor or conductive chemicals. Frequently, however, there is no effective way to ground nonconductive items.

For this reason, air ionizers are a major means of neutralizing static charges. Note, however, that ionizers do not eliminate the need for either wrist straps or protective work surfaces.[3]

Air ionizers generate a constant stream of positive and negative air ions. The positive ions are attracted to negatively charged bodies and the negative ions to positively charged bodies. The result is charge neutralization. The attraction continues until the electric field is neutralized. At that time the ionizer will still be producing ions; the unneeded or unused ions will recombine with other ions.

As you may recall, air ionization occurs when the air receives sufficient energy from induced charge, high voltage ac or dc, or radioactive emissions. When this occurs, an electron is dislodged from an atom or molecule, leaving the atom or molecule with a net positive charge. The free electron will soon latch onto a neutral air molecule and thereby form a *negative* ion. The atom or molecule that lost the electron is called a *positive* ion.[5]

The output of air ionizers should contain nearly equal amounts of positive and negative ions so as to dissipate both the negative and positive charges produced when static electricity is generated. If there is an imbalance of positive and negative ions, residual voltages can develop over the ionized area.[2]

Three types of air ionizers are commonly used: static comb, electric (see Fig. 5-5), and radioactive.

The static comb ionizer is a simple inexpensive unit that requires no power supply. It consists of either grounded needles or metallic brushes placed near the surface to be neutralized. The charged surface creates a potential gradient between the grounded needles and itself. When the voltage becomes high enough, the air will ionize and create a conductive path to ground through the needles.[6] Unfortunately, if the charge is not at least 2,500 volts, it will not operate.[7] As we have noted earlier, electrostatic voltages less than 2,500 volts can damage semiconductors. Therefore, these passive ionizers are used primarily to reduce charging in plastic and paper processing plants.

Electrically powered ionizers consist of two basic components: a high voltage power supply and an ion producing section which is generally a series of needles mounted on a bar. The high voltage power source may supply either direct current or alternating current to the needles. Ac will alternately produce positive and negative ions. For ac ionizers, the power supply may be based on iron-core transformer (ICT) technology or solid state.

ICT power supplies may produce as much as 20,000 volts generally at a frequency of 60 Hz. Because ionization occurs only

Fig. 5-5. Heated air ionizer blower (courtesy of Static Control Systems/3M).

at the peaks of these waveforms, there are several microseconds between the peaks when no ions are being produced. Unfortunately, this may be sufficient time for ESD to occur.

Because of the high voltages present, ozone can be produced as an undesired byproduct, but fortunately this occurs so rarely that it is not a major consideration. Under worst case conditions, ozone can be harmful, as indicated in the following paragraphs.

Ozone (triatomic oxygen) is an unstable colorless or pale blue gas with a pungent noxious odor. It has numerous industrial uses such as purification of drinking water, deodorization of air and sewage gases, and as an oxidizing agent in several chemical processes. In high concentrations ozone is extremely flammable and in liquid form it becomes a dangerous explosive. In low concentrations it can destroy cloth fibers and cause rubber insulation on electrical conductors to crack.

Inhalation of a large quantity of ozone can cause sufficient irritation of the lungs to result in pulmonary edema. Severe exposure may produce vertigo, lowered blood pressure, headaches, and chest pain. Therefore, ventilation should be sufficient to maintain the atmospheric concentration of ozone below 0.1 part per million, which is the maximum level allowed by the Occupational Safety and Health Act.[8]

Solid-state power supplies are said to avoid the problem of ozone by operating at much lower voltages (3000 volts). Operation at reduced levels is possible through control of frequency, amplitude, and pulse width. The frequency of such units is between 15,000 and 40,000 Hz, in contrast to the 60 or 120 Hz produced by iron-core transformer supplies. Because of this higher frequency, lower voltage, and pulse width control, these units are said to produce much less electromagnetic interference (EMI) than produced by ICT power supplies.[9]

The power from the ICT or solid-state power supply is sent to a group of evenly spaced needles held rigidly less than an inch from a grounded metal housing.[5] The charge supplied to the needles will concentrate on the tips of the needles as that is the surface with the smallest radius of curvature. When the charge becomes high enough, the air surrounding the needle points will become ionized.

When the neutralizing needles are connected directly to the power supply, as shown in Fig. 5-6A, the result is a *hot* bar. During operation if any of these needles is accidentally shorted to ground, the entire bar will become disabled as the supply voltage will drop instantly.[5]

An additional problem with hot bars is that a person may accidentally touch one of the needles and get shocked. The type of shock received depends on the power supply. If the power supply is

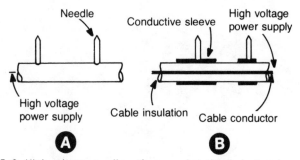

Fig. 5-6. High voltage coupling of powered static eliminators (presented at NEPCON WEST 1981, Anaheim, CA used by permission)[5].

designed, as some, to draw no more than 5 milliamperes, the person will receive an unpleasant, but not fatal, shock. If the power supply does not have this limitation, the shock could be fatal, depending on the current available, resistance of the person, etc. Because of this problem, hot bars are usually placed only where a person is not likely to touch them. When they can be used safely, they are more effective and cost a little less than shockless bars.[10]

In *cold* or shockless bars, the needles are capacitively coupled to the power supply. One way to accomplish this is to imbed each needle in a conductive sleeve surrounding the high voltage transmission cable, as shown in Fig. 5-6B. In this arrangement, the insulator is the dielectric of a capacitor, the cable conductor is one plate of the capacitor, and the outer sleeve is the other plate.[5]

As each needle is an independent circuit, one or more of the needles can be grounded with no effect on the remainder of the assembly.[7] Typically, less than 40 microamperes short circuit current is available at any point so accidental contact is not painful.[10]

The sharper the needle points the more efficient will be the ionizers. However, the high voltage will eventually dull the needle points and the ionizer will lose efficiency. Corrosion and dust can also reduce ion production.[5]

In many situations the ionizer may be several inches or feet away from a charged object. Because of this, fans are added to the ionizers to blow the ions to the charged objects. High air flow is important in order to quickly neutralize the electric fields but if it is too high it may blow items off the workbench and be excessively noisy.

Two types of ionized air blowers are available: bladed fan and squirrel cage. It has been shown[5] that these types have very

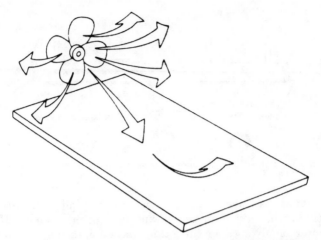

Fig. 5-7. Bladed fan air flow pattern (presented at NEPCON WEST 1981, Anaheim, CA used by permission)[5].

different flow patterns (see Figs. 5-7 and 5-8). As shown, the bladed fan gives incomplete coverage for a typical work area. The squirrel cage blowers give better coverage with less turbulence.

Depending on the amount of charge and the distance of the charge from the ionizing source, ionizers can take several seconds or even minutes to dissipate charges. For this reason, ionizers should be turned on for at least 2 to 3 minutes before commencing work, to allow charges in the area to be neutralized.[2]

Some ionizers do not give a proper balance of positive and negative ions. They leave residual voltages high enough to

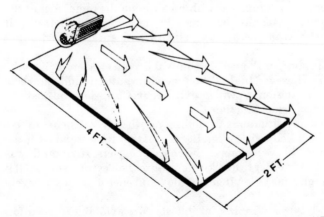

Fig. 5-8. Squirrel cage blower air flow pattern (presented at NEPCON WEST 1981, Anaheim, CA used by permission)[5].

damage some sensitive ESDS items. It is important, therefore, in selecting and placing ionizers to measure the residual voltages in the area and to compare them with the voltage sensitivity levels of the ESDS items being handled.[2]

Ionizers do not give permanent protection for static producing materials. Once these materials are removed from the ionizer's field, they will again produce electrostatic charges.

In addition to the problem of production of ozone, ionizers sometimes have these problems: in the vicinity of high voltage equipment the ionized air may cause inadvertent high voltage breakdown;[3] in atmospheres containing volatile solvents an ionizer may cause a fire or explosion;[6] may cause electromagnetic interference from the blower as well as the ionizing process;[5] and generate untolerable noise for nearby workers.[5]

5.4.1 Radioactive (Nuclear) Static Eliminators

Nuclear static eliminators use a radioisotope such as polonium-210 (or the less common americium-241) to neutralize static charges. These radioisotopes emit alpha particles which are the equivalent of a helium nucleus (two protons, two neutrons). As these particles streak through the air, they collide with air molecules and thereby create positive and negative ions.

The alpha particles emitted by these radioisotopes have high energy but a low penetrating ability; in fact, they will not penetrate a thin sheet of paper or the epidermal layer of skin. In air they have a range of less than 2 inches. In some cases they do not even leave the housing of the eliminator even though they still continue to ionize the air. Despite the fear of some people of any source of nuclear radiation, no matter how harmless, there appears to be no external hazard associated with the use of such devices.

As typical of radioactive devices, these units lose their effectiveness with time. The radioisotope is said to "decay." Because of this decay, they have a normal useful life of one year at which time the nuclear element must be replaced by the manufacturer. Thus the units must be leased rather than purchased from the manufacturer.

The eliminators are available in two basic shapes: bars and small circles. Often they are attached to blowers in order to protect a larger area.

Because they are not electrically powered, they cannot generate electric sparks which could cause explosive atmospheres to ignite. They can be easily moved from one work station to another since they require no external power supply and wiring.

5.4.2 Ionizing Air Guns

Compressed air or nitrogen is frequently used to remove dust and other contaminants from objects being manufactured or assembled. If these objects have a static charge, however, they will again attract dust once the compressed air is removed. Worse still, the air may create additional charges on the objects.

To eliminate this problem, ionizers have been added to compressed air guns to neutralize static charges while blowing away dust. The gun itself is a hand-held, pistol shaped device. It is connected by several feet of shielded cable to a power unit. A filter in the gun removes oil, moisture, and dust from the compressed air. Ionization is provided by either radioactive emission or high voltage electrical discharge.

Obviously, the gun must be shockproof in the case of electrical discharge. To minimize fatigue, it needs to be lightweight. One manufacturer incorporates an ultrasonic generator in an air gun to aid in removing small particles. A variation of the ionizing air gun is the ionizing air nozzle which supplies a narrow stream of ionized air at a fixed point on a continuous basis.

5.5 TOOLS AND PRODUCTION EQUIPMENT

From the smallest shop to the most sophisticated electronic equipment manufacturing operation, there are tools and production equipment that can add to the ESD problem.

Soldering irons, solder pots, or flow soldering equipment should be hard grounded and isolated from the power line by transformer or direct current. To keep the voltage buildup of a hot soldering iron less than 15 volts, the resistance between the tip of the iron and ground should be less than 20 ohms. Other electrical power equipment which could contact ESDS items should also be grounded. ESD protective solder suckers should be used. Periodically, the insulated handles of hand tools should be checked for static generation and treated with an antistat (see section 5.6) if necessary. Frequently handled small hand tools often accumulate skin moisture which may make them ESD protective.[2] Conductive work stations, as shown in Fig. 5-9, allow PCBs to be handled safely.

Grounded baffles are necessary for temperature chambers to dissipate charges in circulated air. Instead of baffles, ionized air can be used in the chamber to dissipate static charges caused by air flow, or shields can be used to divert the charged air away from ESDS items in the chamber. CO_2 can be used in cooling chambers but caution is necessary because the evaporation of the CO_2 can generate high static charges. The parts tested in temperature chambers should be placed in ESD protective tote boxes or trays on grounded metal racks within the chamber.[2]

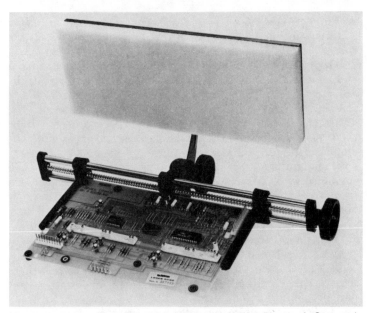

Fig. 5-9. Conductive work station (courtesy of Micro Electronic Systems).

When spraying, cleaning, painting, or sandblasting of ESDS items occurs, ionized air blowers, conductive solvents, or ionized nozzles should be used as applicable to prevent electrostatic charge buildup in the work area. Wet blast conductive or antistatically treated slurry with a maximum volume resistivity of 500 ohms per cm should be used instead of dry sandblasting. Low resistivity solvents such as ethanol mixed with a normal cleaning solvent give improved control of charge generation.[2]

5.6 TOPICAL ANTISTATS[2]

Topical antistats are chemical agents, which when applied to the surface of insulative materials will reduce their ability to generate static. This is accomplished in two different ways. First, they increase the *surface lubricity,* thereby reducing the material's coefficient of friction. With less friction on the surface, triboelectric charging is reduced. In addition, topical antistats increase *surface conductivity* which allows charges to bleed off.[3]

Charge dissipation by surface conductivity can be explained by two different theories. One theory says that the increased conductivity allows the antistat and the material it covers to exchange enough electrons to maintain an electron balance. The other theory is that the antistat introduces a balance of positive and negative ions on the material's surface. There are enough ions to neutralize most of any charge generated on the surface. In addition, an ion exchange with the air further dissipates the charge.[11]

Topical antistats are generally liquids consisting of a carrier and an antistat. The carrier is the vehicle used to transport the antistat to a material. It acts as a solvent and can be water, alcohol, mineral spirits, or other compatible material. The antistat is the substance that remains deposited on the material's surface after the carrier evaporates and it provides the static control function.

Some antistats are detergent type materials, which when combined with the moisture in the air wet the surface on which they are deposited. The effectiveness of these hygroscopic antistats is reduced under low relative humidity.

Topical antistats are applied by brushing, spraying, rolling, dipping, mopping, or wiping. They can be applied to floors, carpets, workbench tops, parts trays, parts carriers, chairs, walls, ceilings, tools, paper, plastics, and clothing to give them some degree of ESD protection. Once treated, an object does not normally require grounding.

Topical antistats are particularly useful in cleaning work surfaces and floors and in treating exposed plastic surfaces, which cannot be effectively treated by other ESD control techniques.[3] They should not be applied to electrical circuit boards, parts, or assemblies because they can create leakage paths and possibly affect solderability.

During cleaning operations topical antistats can be unintentionally removed, thereby requiring the cleaned surface to be retreated with antistats. When not removed by cleaning, topical antistats can provide protection for a long time, depending on the application rate and the amount of handling.

Because antistats do not give permanent protection, their effectiveness should be periodically checked by either of two methods: (1) by rubbing the treated area with common polyethylene and monitoring the charge and its decay time with an electrostatic field meter, or (2) by measuring the surface resistivity of a sample of the material using the appropriate test equipment.

Items made of ESD protective materials that require periodic treatment with a topical antistat should have a sticker showing the date that the ESD protective properties of these items should be rechecked. Such items could include tote boxes, trays, and gauntlets.

In selecting an antistat, numerous characteristics must be considered in addition to its antistatic properties:

1. Ability to inhibit bacterial growth (antistats which can do this are called *bacteriostatic*)
2. Nontoxicity
3. Noncorrosivity
4. Nonflammability
5. Nonirritating to personnel (including eyes)
6. Nonstaining

Additional factors suggested by the Reliability Analysis Center[3] are:

1. Longevity and wear characteristics
2. Decay performance and controllability
3. Ease of application
4. Performance compatibility with your materials.

6

Protected Worker

A S WE HAVE SEEN IN PREVIOUS CHAPTERS, ELECTROSTATIC DISCHARGE sensitive devices need to be kept in protective packages whenever possible. Outside such packages they need to be in a protective environment. In addition, such devices should be handled only by protected personnel. This protection is provided primarily by grounded wrist straps and secondarily by special clothing.

6.1 GROUNDED WRIST STRAPS

Wrist straps are probably the most common type of defense against ESD because they are inexpensive yet effective. They provide the critical first line of protection. By maintaining an electrical path between ground and the worker, they eliminate (or dramatically reduce) any voltages that might be created on an ungrounded worker. Figure 6-1 shows typical voltages on an ungrounded person. In contrast, a person wearing a grounded wrist strap will have a peak voltage no more than 11 volts.[1] It's obvious, then, that if a worker forgets to attach his wrist strap to ground, he can do tremendous damage to ESDS devices.

A wrist strap consists of a cuff (the piece that encircles the wrist) a lead (the wire or conductive fiber connecting the cuff to ground) a series resistor (a resistance in series with the cuff and the ground) and various connections (see Figs. 6-2 and 6-3).

The two basic types of cuffs are:

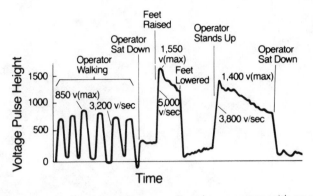

Fig. 6-1. Typical voltage variations monitored on a person with no wrist strap (courtesy of J. R. Huntsman[1]).

- Carbon-impregnated plastic straps that incorporate their own built-in resistance. This type should be insulated over its length to prevent it from touching a hard ground and thereby provide a low resistance to ground.
- Metal conductor straps, which are insulated on the exterior side and use a series resistor.

Cuffs are frequently held together by a strip of Velcro, a material which is easily connected or disconnected. When joined, none of the Velcro should be exposed because it might snag and damage components or clothing.[2]

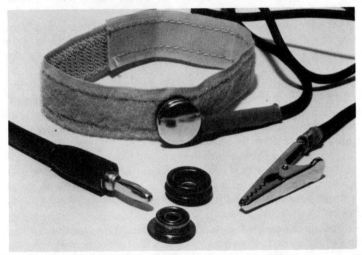

Fig. 6-2. Wrist strap (courtesy of Charleswater Products).

Fig. 6-3. Wrist strap in operation (courtesy of Static Control Systems/3M).

It's important that the cuff make good skin contact so as to rapidly discharge personal static charge to ground and equalize the person's static level with that of the work surface. To obtain this snug skin contact, the strap should be adjustable and be placed directly against the skin with no clothing between the two.

The series resistor is necessary to prevent wrist straps from becoming a personnel safety hazard. If, for example, this resistance is 250,000 ohms, it will protect personnel from shocks up to 1,250 volts by limiting the current to 5 milliamperes. A 1-megohm resistor is often used because in most cases it would limit the current to 1 milliampere, which is easily tolerated. Smaller values of resistances can be used if it can be ensured that the voltages on operator could possibly touch will not cause more than 5 mA to flow through him or her.

By keeping the resistance to ground to a minimum, residual voltages on people performing static generating motions will be minimized. For handling extremely sensitive parts such as gallium arsenide diodes it may be necessary to reduce the series resistor to 100 kilohms.[3]

Some engineers feel that the location of the series resistor in the ground lead is important. Typically it is encapsulated in a molded snap housing at the cuff end of the lead. If the resistor were at the other end of the lead, the lead could conceivably short to ground ahead of the resistor, thereby shunting the resistance. Strain relief at the wrist and ground connections give longer life to the wrist strap.

Wrist straps should not have any exposed metal or conductive plastic parts if they are to be used near exposed or open voltages. In most, but not all, assembly areas this may be no big concern as there are usually no easily accessible voltage points. It may be a problem, however, in test areas during troubleshooting or repair of live equipment.

To equalize the potentials between the worker and the workbench top, the wrist strap should be connected to the bench top at a common terminal where the workbench cable is also connected.[4] This connection should be an alligator clip or other quick-release mechanism so that the strap can be easily released in emergencies without injury to the person. For the same reason the strap should have an easy release connection at the point of contact at the cuff. But neither release should be so easy that the worker would not be aware that the connection was broken. In addition to protecting the worker during emergencies, the quick-release connections protect the worker from injury in case he or she forgets to disconnect the strap before walking away from the work area.

In the selection of wrist straps, operator acceptance is an important consideration because the operator is not likely to use it 100 percent of the time if he or she doesn't like it.[2] This acceptance is based on subjective matters such as comfort, ease in putting on and removing, and attractiveness.

Whether sitting or standing, a worker should wear a wrist strap the entire time he or she is at the work station. The strap should be worn on the left wrist for a right-handed person, on the right for a left-handed person.[5] Some manufacturers recommend that wrist straps be worn when personnel remove or insert printed circuit boards in equipment. In board test areas straps may be impractical because workers do not stay in one position.

Even when wearing a wrist strap, a worker should first touch the grounded bench top before handling ESDS items. And even though grounded, a worker should avoid touching leads or contacts.[5]

It is obvious, perhaps, that the worker needs a wrist strap. It may not be so obvious, however, that *any* visitor to that person's work station needs a wrist strap, too, if he or she is going to touch any item on the workbench. Field engineers should carry a wrist strap with them for use during equipment repair and adjustment.

Because wrist straps wear out from frequent use, they should be checked daily (some say several times daily) and at the time of receipt to ensure they are operating properly and have no open connections. A tester is available that compares the strap's resistance with one of three standards built into the unit: 270 kilohms, 510 kilohms, and 1 megohm.

The wrist strap system tester shown in Fig. 6-4 checks not only the continuity of the strap, but also the integrity of the strap protection resistor, satisfactory strap-to-skin contact resistance, and a proper ground connection. To encourage frequent checking, such testers should be placed at each work station.

Wrist straps provide protection only against static charges on the body; they do not help alleviate static charges on the clothes the person is wearing. The following section discusses the role of ESD protective clothing.

6.2 ESD PROTECTIVE CLOTHING

Because the human body is a conductor, it may be effectively grounded. Clothing, however, will generate and retain electrostatic charges, sometimes as high as 30 kilovolts (in the case of synthetics). Unfortunately, these static charges cannot be effectively drained or dissipated by conventional grounding.[6]

In an area where ESDS items are handled, a person's clothes may be significant insofar as ESD control is concerned. By clothing, we mean any garment or apparel on a person's body. This includes shoes, underclothes, and outer garments including lab coats. In clean rooms it includes these garments as well as coveralls, head covers, gloves, and finger cots. Outside clean rooms, it is difficult to dictate and then enforce the wearing of clothes that do not produce ESD, but, nevertheless, it should be suggested.

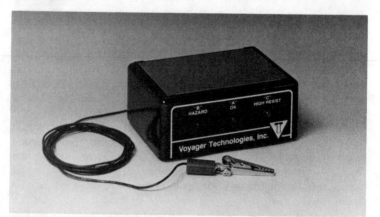

Fig. 6-4. Wrist strap system tester (courtesy of Voyager Technologies. Inc.).

Fig. 6-5. Neutro-Stat antistatic garments (courtesy of SIMCO).

In any work environment workers should be especially careful to prevent any ESDS items from touching their clothing. Therefore, short-sleeve shirts or blouses are preferred. If long sleeve clothes are worn, the sleeves must be rolled up or else covered with an antistatic sleeve protector, called a gauntlet, from the bare wrist as far up as the elbow.[5] When gauntlets are worn, the clothing does not have to be made of ESD protective materials.[4]

Antistatic garments can be made by treating cotton or synthetic clothes with an antistatic chemical agent in the final rinse during laundering. Each time these clothes are washed, the antistatic agent must be reapplied.[6]

Another common type of antistatic garment is made from a fabric of 65 percent polyester, 34 percent cotton, and 1 percent stainless steel fiber. See Fig. 6-5. The steel fiber is an integral part of the fabric. Surprisingly, the fabric can be worn, handled, and laundered just the same as ordinary clothing.[6]

In critical cleanroom applications, however, neither type of garment is acceptable. Cotton garments, for example, will shed too much lint, even if they are blended with synthetic fabrics. The steel fiber breaks up into fine particles during extended processing and use.[7]

Nylon and Dacron garments have been widely used in clean rooms because they are essentially lint free. At the same time, however, they produce tremendous static charges. By treating them with antistatic agents each time they are laundered, these garments can be used, but they are far from ideal, as shown in

Table 6-1. Antistatic Treatment Techniques
for Woven and Nonwoven Fabrics (Courtesy of E. S. Burnett[7])

Technique	Advantages	Disadvantages
Disposable garments (Tyvek)®	1) Good to 20% R.H. 2) Very low initial cost	1) Particulates? 2) Low abrasion Resistance 3) Cannot currently be reused in low R.H. environment
Dacron® with woven nylon conductive fiber	Permanent	1) Higher initial cost 2) Acid erosion a minor problem
Dacron with standard topical treatments	Low Cost	Fails to meet specs. at lower humidities.
Dacron with experimental topical treatments	Meets specifications at low humidity	Currently very costly
Spray	Not developed as a control technique	Uneven distribution

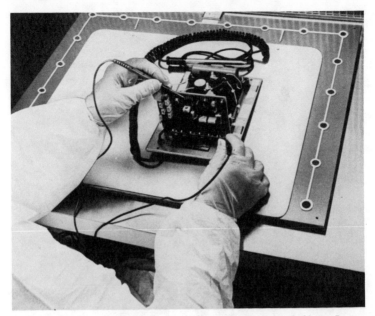

Fig. 6-6. Static-barrier clean room gloves (courtesy of Clean Room Products).

Table 6-1. Tyvek® garments show promise as being the best compromise, but additional investigation is needed to confirm this.

Gloves or finger cots are used in clean rooms to keep body moisture, oils, and dust off sensitive items. Either, however, can cause ESD problems. A new powder-free glove (Fig. 6-6) has been designed that is claimed to be static free. These gloves, it is claimed, eliminate the need for wrist straps. They are said to be tough, yet ultra-thin, and to have a minute gap between glove surface and the hand for breathing and vapor dispersal. Where finger cots are used, they should be made of ESD protective materials. It is also recommended that they be used only in an ionized-air environment.[5]

Shoe soles can be leather, composite, crepe, or rubber. Because of its much lower resistance, leather is recommended while crepe and rubber soles are taboo. To insure proper contact to ground is maintained regardless of footwear, disposable electrically conductive shoe grounding straps (Fig. 6-7) can be used. Shoe straps are particularly valuable in areas where personnel must move about unhindered by wrist straps. Conductive shoes and shoe covers should be tested when initially put

Fig. 6-7. Pressure sensitive conductive shoe strap (circled) (courtesy of Static Control Systems/3M).

on and when re-entering an ESD controlled area. They should not be worn outside the building.

Protective clothing should be frequently checked (especially after cleaning) with an electrostatic field meter to monitor for damaging ESD voltages. High readings will be found at material edges: creases, cuffs, and hems. The highest readings will be found on workers wearing multiple layers of synthetic materials.[8]

7

The Complete
ESD Control Program

I N PREVIOUS CHAPTERS WE HAVE LOOKED AT THE CAUSES OF ESD AND ITS effects on electrical and electronic parts and at ESD protective materials and equipment. For the complete ESD control program, however, we need to consider ESD control program guides, the design and construction of ESD protected areas, the preparation of ESD operating, handling, packaging, and marking procedures, the development of ESD personnel training programs, and certification of ESD protected areas and grounded workbenches.

7.1 SETTING UP AND MONITORING AN ESD CONTROL PROGRAM

To implement a successful ESD control program, it is absolutely essential to have the support of top management. If these people are convinced of the need for ESD protective measures, then they will be willing to authorize the purchase of ESD protective materials and to direct the necessary changes in component handling procedures.

Whether this is a top management decision or an intermediate management decision may be determined by the costs of implementing the program and by the number of departments involved in resolving the ESD problem. Quality assurance and engineering will certainly be affected, but so will manufacturing, field engineering, and numerous other departments. To leave any pertinent department out of the planning and implementation could obviously be disastrous.

Fig. 7-1. Static safe-guarded work station (courtesy of Static Control Systems/3M).

Once top management approves the concept of ESD control, an ESD task force or committee of all the key elements in the company should be established to implement the program. This committee's goal should be to "ensure that products are designed, produced, handled, and maintained in a way that achieves the highest ESD immunity within sound economic means."[1]

Typical objectives for this committee are:

1. Implement a design specification that gives the minimum acceptable levels of product susceptibility to ESD.

2. Raise employees' level of awareness of the impact of ESD on the company's business.
3. Implement manufacturing processes that control and minimize the effects of ESD.[1]

After the program is set up, it must be monitored to ensure that all personnel are continuing to follow the proper ESD control procedures.

7.2 PROTECTED AREAS[2]

The key to a successful ESD control program is a protected work area (Fig. 7-1) which includes an ESD grounded workbench. With such an area, whenever an ESDS item is handled outside of its ESD protective packaging, electrostatic voltages will be kept below the sensitivity level of ESDS items. The lower the level of the static generated voltage, below the ESDS item sensitivity level, the greater is the probability of protecting that item.

The degree of sophistication needed for a protected area depends upon the level of assembly and maintenance (for example, equipment or assembly, organizational, intermediate or depot) and on physical limitations of the work area or facility. For example, for field maintenance a protected area could

Fig. 7-2. Portable field service kit (courtesy of Static Control Systems/3M). Systems/3M).

111

Table 7-1. Model Specification
for ESD Protection (Courtesy of L. P. Phillips[3]).

Item	Remarks
1. Work Area a) Size: 2000-3000 sq. ft.	i Accommodate 30-50 people. ii Limit the number of exits and entrances, hence environmental conditons are controllable
b) Floors: Conductive tiles	i Easier to maintain and clean.
c) Ceiling and ventilation	i Necessary for stability of environmental control.
d) Lighting level	i 300 ft. candles.
2. Environmental Controls	
a) Humidity: 45% ± 5%	i Comfortable and pleasant.
b) Temperature: 68°F ± 5°F	i Heat generating equipment will raise the temperature to 76°F (24.4°C) ii Limits rusting and corrosion. Reduces ESD arc generation.
c) Smoking	i Smoking should not be allowed. ii Smoke particles can form discharge paths or deposit charges when they settle on assemblies in work areas. iii Nicotine and other contaminants create poor solderability and conformal coat problems.
d) Food and beverages	i No eating or drinking is to be allowed. ii Fruit juices will encourage bacterial or fungal growth, contaminate PC boards and assemblies.
3. Bench Top Materials	
a) Surface resistivity range $10^6 - 10^9$ ohms per square	i Material manufacturers should supply data sheets that specify resistivities at RH levels of 30%-60% and temperatures of 60°F-80°F.
b) Static decay times 0.1 - 0.2 seconds from 5000 volts.	ii Very fast decay times generate secondary RF fields that can destroy ESDS devices via discharge to highly conductive containers (metallic)
c) Chemical properties	iii Must not support bacterial or fungal growth. iv Must not form ionic layers with PC boards to release particles and other contaminants from solderable surface.

Item	Remarks
3. Bench Top Materials (cont'd) c) Chemical properties (cont'd)	v Shall be nonflammable, nontoxic by oral injection, inhalation or dermal application. vi Should provide some cushioning and must not depend on humidity to function.
4. Protective Clothing a) Coveralls, labcoats and smocks	i Cotton labcoats are comfortable. ii Maintenance of low residual ESD voltage on synthetic labcoats is almost impossible. iii Labcoats must not be worn outside static free work areas (i.e. washrooms, cafeterias and outdoors, etc.)
b) Wrist straps	Mandatory
c) Conductive shoes	Necessary in mircocircuits labs or ESD protected areas with conductive floor.

5. Protective Material
 	
   a) Tote boxes, containers
 	
   b) Transport carriers
 	
   c) Document holders
	Should have the same properties as per paragraph 3 of Model Specification.

6. Ionizers	Use with caution especially in conjunction with chemicals such as conformal coat areas.
7. Cleaning agents Antistatic sprays and solutions	i Must have same specifications as per paragraph 3c. ii Must not be corrosive. iii For microcircuits labs and fabrication areas, acid radicals, sodium, phosphates, chlorides and sulphates-free solutions must be used.

consist of an area kept free from static generators and equipped with a personnel wrist ground strap, a portable protective work mat, ESD protective packages for repair parts and spares, and ESD protected hand tools such as grounded-tip soldering irons. A typical portable field service kit is shown in Fig. 7-2.

On the other hand, a protected area in a manufacturing facility would include humidity controls, an elaborate grounded workbench made of ESD protective materials, personnel wrist ground straps, wrist strap checker, grounded tools and equipment, air ionizers, and other protective equipment. Table 7-1 gives specifications for an ideal ESD protective area.

Tradeoffs should be performed to determine the proper balance between the cost of constructing protected areas versus the complexity of the handling procedures to be implemented to get the required controls. Training and controls, for example, should consider the ability and willingness of people to follow elaborate handling procedures and to use certain ESD protective materials and equipment such as wrist straps and ankle straps.

Two approaches are common to the design of ESD protected areas: (1) consider all ESDS items as highly sensitive and then standardize controls throughout the facility, or (2) implement only those controls needed for the sensitivity of the items pertaining to a specific protected area.

One of the basic principles in the design of protected areas is to prohibit the use of prime generators (Table 2-2) and to restrict the entry of these prime generators by personnel working in these areas. That is, plastic drinking cups, notebooks, purses, and paper covers must not be allowed in the immediate work area.

Protected work areas, workbenches, and containers should be identified by ESD caution signs such as "ESD Protected Area: Use Precautions When Handling ESDS Items Outside of Their Protective Wraps."

Access to protected areas should be restricted to people who are properly trained and attired, or who are escorted, cautioned about protective procedures, and restricted from contacting ESDS items.

7.3 GENERAL GUIDELINES FOR HANDLING ESDS ITEMS[2]

The following general guidelines apply to the handling of ESDS items:

1. People handling ESDS items should be trained in ESD precautionary procedures, tested for competency, and certified. Untrained personnel should not be allowed to handle ESDS items when the items are outside of the ESD protective packaging.

2. When not actively working with ESDS items, using the terminals for testing, or inserting the terminals of a part in a printed wiring board or assembly electrical socket, workers should use shunts such as bars, clips, or noncorrosive conductive foam or protective covering to protect the item from triboelectric generation, direct discharge, electrostatic fields, and electromagnetic pulses from high voltage ESD spark discharge. When inserting part leads into printed wiring boards terminal holes or making connections to part leads, workers should insert shorting bars, clips, or conductive foam over the connector terminals at the printed wiring board or higher assembly level. For protection from electrostatic fields and electromagnetic pulses from high voltage ESD spark discharge, conductive outer wrapping should be used, especially when the item is being transported outside of ESD protected areas.

3. People maintaining ESDS equipment where personnel ground straps cannot be used should ground themselves prior to removing ESDS items from their protective packaging. When being handled out of their protective packaging, ESDS items should be handled by the shunting device, without touching ESDS parts or electrical runs.

4. The leads or connector terminals of ESDS items should not be probed by multimeters (VOM). If this is not practical, personnel should touch ground with the electrical test equipment probes before probing the ESDS item.

5. Tools and test equipment used in ESD protective areas should be properly grounded; hand tools should not contain insulation on the handles or, if used, tools with insulated handles should be treated with a topical antistat.

6. Power should not be applied to equipment or assemblies while ESDS items, especially those containing MOS devices, are being removed or inserted. Additional precautions for MOS are as follows:

   a. Signals should not be applied to the inputs while MOS device power is off.

   b. When MOS devices are being tested, all unused input leads should be connected to either source ground or V_{SS} or supply, V_{DD}, whichever is appropriate for the circuit involved.

   c. Prior to performance of dielectric or insulation resistance tests, MOS should be removed from the equipment.

    d. All power supplies used for testing ESDS items should be checked to ensure there are no voltage transients present.

    e. Test equipment should be checked for proper voltage polarity before conducting parameter or functional testing.

7. All containers, tools, test equipment, and fixtures used in ESD protective areas must be grounded before and during use, either directly or by contact with a grounded surface. Grounding of electrical test equipment should be via a grounded plug, not through the conductive surface of the ESD grounded work station.

8. Carriers, holders, or containers should be electrically connected before transferring ESDS parts from one to the other.

9. Work instructions, test procedures, drawings, and similar documents used in ESD protected areas should not be covered in common plastic sheeting or containers.

10. Drawings for ESDS items should be marked with the ESD sensitive electronic symbol, warnings, precautions, and handling procedures as applicable.

11. Manufacturing, process, assembly, and inspection work instructions should identify ESDS items as needed for ESDS control and require that such items be handled out of their ESD protective packaging only by trained personnel and only in ESD protected areas.

12. Personnel handling ESDS items should avoid physical activities that are static producing in the vicinity of ESDS items. Such activities include wiping feet and removing or putting on smocks.

13. Personnel handling ESDS items should wear ESD protective clothing. Such clothing should be monitored periodically with static meters. Close-fitting, short sleeve "street" garments are acceptable for ESD purposes provided they do not contact ESD items directly. Common synthetic clothing should be regarded as a static hazard and work habits should be developed to prevent its contact with ESDS items. Gloves and finger cots, if used, should be made of ESD protective material.

14. Periodic continuity and resistivity checks of personnel ground straps between skin contact point and ground connection, ESD grounded workbench surfaces, conductive floor mats and other connections to ground should be performed with suitable test equipment to

ensure conformance with grounding resistivity requirements.

15. Charges of ESD protective packaging containing an ESDS item should be neutralized by placing the packaged item on an ESD grounded workbench surface to remove any charge prior to opening the packaging material. Charges can also be removed by grounded personnel touching the package.

16. ESDS items should be removed from ESD protective packaging using finger or metal grasping tool only after grounding and then placed on the ESD grounded workbench surface.

17. When ESDS items are being tested in test chambers using carbon dioxide or nitrogen gas for cooling, the chamber should be equipped with grounded baffles and shelves, on which equipment rests, that are grounded to dissipate electrostatic charges created by the flow of the gas. In-line ionizers for the gas are needed when the chambers or ESDS items have insulating surfaces.

18. When ESDS items are manually cleaned with brushes, only brushes with natural bristles should be used and ionized air should be directed over the cleaning area to dissipate any static charges. All automatic cleaning equipment should be grounded if practicable and leads and connectors of ESDS items should be shorted together during the cleaning operation. Conductive cleaning solvents should be used where practicable when cleaning ESDS items.

19. Caution should be observed in using solvents such as acetone and alcohol or other cleaning agents for cleaning ESD protective materials. The use of such solvents can reduce the effectiveness of some ESD protective materials, especially those employing detergent type antistats which bleed to the surface to form a sweat layer with moisture in the air.

20. The cases or chassis grounds of test equipment and ESDS items being tested should be electrically connected together prior to connecting or disconnecting any test cables. When connecting test cables, shunting bars should remain in place until chassis grounds are shorted. Shunting bars should be replaced upon removal of test cables.

7.4 SPECIFIC AREA HANDLING PROCEDURES[2]

The preceding guidelines are useful for all areas in a plant. The following procedures are needed for specific areas as indicated.

7.4.1 Receiving Inspection

Procedures are as follows:

1. Be aware of all ESDS items to be delivered by vendors.
2. Remove the unit package of an ESDS item from the shipping container. Do not open the unit package. Examine the item for proper labeling and ESD protective packaging in accordance with the procedure or contract. Inspect as follows:
   a. Labeled packages—examine packages of ESDS items to verify conformance to the precautionary labeling and ESD protective packaging requirements of the contract.
   b. Non-ESDS marked packages—while the supplier is responsible for proper marking of packages containing ESDS items, the treatment of such items delivered without an ESDS marking should be governed as follows:

■ No ESDS marking, but in protective packaging—the packaging should be marked with proper markings and these ESDS items should be handled in accordance with the procedures of this book. The supplier should be reminded to ensure proper marking of future shipments.
■ No ESD marking and no ESD protective packaging—the ESDS item should be rejected as defective and should not be accepted if resubmitted by the supplier.

   c. Open the unit packaging and perform tests of ESDS items only in a protected area and if size permits at an ESD grounded workbench. Follow the general guidelines given previously.
   d. Repackage tested ESDS items in ESD protective packaging material and ensure proper marking on the package.
   e. Place unit packages in ESD protective tote boxes or trays for transporting to stores.

7.4.2 In-Process Inspection and Test

Procedures are as follows:

1. Observe the general guidelines given previously.
2. Open unit packaging of ESDS items only in protected areas at grounded workbenches.
3. After examination and test, repackage ESDS assemblies in ESD protective packaging materials and ensure proper marking on unit packaging. Place ESDS as-

semblies in protective tote boxes or trays for transporting to next workbench.

7.4.3 Material Receiving Area

Procedures are as follows:

1. Be aware of all ESDS items to be delivered by vendors.
2. Do not open packages containing ESDS items. Peform quantity counts of ESDS items to compare with contract qualities. If protective packaging is opaque or if counts cannot be verified without opening ESD protective packaging, quantity counts should be deferred to Receiving Inspection.

7.4.4 Storeroom Area

Procedures are as follows:

1. Transport ESDS items to and from the stockroom area in ESD protective packaging and tote boxes or trays that will protect the ESDS items from triboelectric charges, discharges from personnel or objects, electrostatic fields and electromagnetic pulses from ESD high-voltage spark discharge.
2. Do not open unit packages of ESDS items for count issuance or kitting unless required. When required, opening of unit packaging of ESDS items should be performed at grounded workbenches observing the general guidelines given previously. Repackage ESDS items in ESD protective packaging and mark packaging in accordance with the MIL-STD-129 sensitive electronic device symbol and caution.
3. Ensure all packages and kits issued from the stockroom containing ESDS items are marked with the ESD sensitive symbol and precautions.
4. Identify ESDS items on all kitting documentation.

7.4.5 Production, Processing, Assembly, Repair, and Rework

Procedures are as follows:

1. Perform operations only in protected areas and where practicable at ESD grounded workbenches.
2. Use static eliminators or ionized air blowers to direct a stream of ionized air over the ESDS item being processed when grounding techniques cannot be used to equalize or dissipate electrostatic charges.
3. Perform cleaning processes using conductive cleaning fluids or solvents. Where solvents are air sprayed, use ionized air.

4. Upon completion of assembly and processing of the ESDS assembly, repackage in ESD protective packaging material and ensure proper marking. Place the ESDS assemblies in protective tote boxes or trays for transporting to next workbench.

7.4.6 System and Equipment Level Test and Maintenance

The following procedures also apply to testing in the field:

1. Perform testing of ESDS items only in ESD protected areas to the extent practicable.
2. Observe the general guidelines given previously.
3. Perform diagnostics to isolate the faulty assembly. Do not use canned coolant for fault isolation.
4. Shut off power to equipment.
5. Prior to touching an ESDS item, attach a personnel ground strap to the wrist and connect the other end to the equipment cabinet or chassis ground. Where personnel ground straps cannot be used, touch the grounded equipment chassis with a hand prior to removing or inserting an ESDS item.
6. Upon removal of the failed ESDS item, package in ESD protective packaging material.
7. Do not probe or test ESDS items with a test equipment lead unless necessary; ideally, this should be done only under controlled conditions such as with ionized air. When such probing is necessary, ground the meter and probes or test leads prior to touching the terminals of the ESDS items.
8. Open the package at the connector end if possible. Remove the ESDS item from the ESD protective packaging and install the item in the equipment. Avoid touching parts, electrical terminals, and circuitry.
9. Perform all other required maintenance actions such as tightening of fasteners, adjustments, and replacement of covers prior to removal of the personnel ground strap. If a ground strap is not worn, touch the grounded equipment chassis prior to each action.
10. Energize equipment and verify correction of fault.

7.4.7 Packaging and Shipping Area

Procedures are as follows:

1. Ensure that all ESDS items submitted for shipment have been received in ESD protective packaging material.

2. Remove items from interim packaging only at an ESD grounded workbench, observing the handling guidelines listed previously.

3. Package the ESDS item in ESD protective material for shipment as required by the contract.

7.5 PERSONNEL TRAINING AND CERTIFICATION

Without properly motivated and trained workers, even the best ESD protected areas and ESD protective handling procedures will not provide the protection needed. Workers can generally be motivated if they are fully convinced that wrist straps and other protective devices and detailed ESD handling procedures are important. Until they are convinced, they are likely to feel that the straps and the procedures are a pain in the neck, degrading, and unreasonable. Even though top management may send down directives, the rules may be ignored, especially if there are no inspectors around, unless the workers see the need to change their work habits.[4]

How can workers be convinced that ESD is their problem? While an aggressive training program will include training films and pertinent lectures, it is generally agreed that the best way to convince a worker is to allow him or her to purposely destroy or zap a device with ESD.[4,5,6,7] If ESD damage is demonstrated at worker's station, he or she is more likely to become a believer.

Training in static awareness and ESDS protective procedures should be provided to all people who specify, procure, design, or handle ESDS items. At Douglas Aircraft the training was extended to everyone who had access to ESDS items at any time, for example, maintenance personnel.[8]

ESD training programs should be oriented to the contractor's facilities and the types of ESD materials and equipment that have been found to be effective for his particular application. Personnel should be trained to effectively employ the ESD protective materials and equipment provided and to understand the theory behind many of the ESD precautions included in ESDS handling procedures.

ESD awareness should also be a part of equipment training courses prepared by contractors for the user. Such training should include identification of ESDS items in the equipment, basic ESD theory, ESDS handling precautions, the need for, use of and types of ESDS protective packaging and the safety aspects involved where grounding is a part of the ESD handling procedures.

The depth of training should depend on the trainee's ability and need to comprehend the information provided. Obviously, engineers who design protective circuitry will need more theoretical training than will stockroom personnel who place ESDS items in kits. Certificates of satisfactory completion should be given those who have attended the training course and demonstrated a comprehension of the course. Periodic re-education of all personnel will be needed as will as some program to ensure that all new personnel receive the appropriate training.

7.6 PRODUCT DESIGN

In designing ESD immunity into a product, it is vital to consider where the product will be used and who will use it. For example, in a computer room or a laboratory, ESD protective materials and procedures can be easily implemented. Therefore, equipment destined for this environment can have a lower ESD immunity than equipment to be used in industrial or public installations. The solution to this requirement is not easy because equipment for public or industrial usage is likely to have the smallest profit margin.[1]

One primary solution is to house electronic equipment in metal boxes, if possible. If not possible, then the outside of the equipment should be a nonconducting material that surrounds an inner metal box. Cables leaving the equipment should be shielded and the shield connected to the chassis ground.[1]

To produce a static-protected product will require close cooperation between various engineering disciplines. Without this cooperation, a mechanical engineer, for example, could design a conventional plastic cabinet which could meet all requirements except static protection.[9]

7.7 CUSTOMER RESPONSIBILITIES

Ideally, a customer should be able to ignore ESD problems. But, until that happy day arrives, the customer must share the responsibility of ESD damage. The customer can do this only if he is made aware of the problem through product manuals and other communication.

Appendix

MANUFACTURERS AND SUPPLIERS

WRIST STRAPS
  Charleswater
  Colvin
  Control Static
  NRD
  Nu-Concept
  Scientific Enterprises
  Semtronics
  SIMCO
  Static Control Systems/3M
  Wescorp

PROTECTIVE WORK SURFACES
  Bengal
  Charleswater
  Plastic
  Richmond Division of Dixico
  Semtronics
  SIMCO
  Static Control Systems/3M
  Vinyl Plastics
  Wescorp

PROTECTIVE BAGS
  Armand
  Bemis

Protective Bags *(continued)*
- Bengal
- Charleswater
- Colvin
- Controlled Static
- Jiffy
- NRD
- Richmond
- Sealed Air
- Semtronics
- SIMCO
- Static Control Systems/3M
- Techni Bag
- Wescorp

CUSHIONING MATERIALS
- Armand
- Bemis
- Bengal
- Charleswater
- Controlled Static
- Du Pont
- Gary
- Jiffy Packaging
- Republic Pack
- Sealed Air
- Semtronics
- Sentinel Foam
- SIMCO
- Static Control Systems/3M
- Techni Bag
- Wescorp

CONDUCTIVE SHUNTS
- Charleswater
- Colvin
- Controlled Static
- NRC
- Semtronics
- SIMCO
- Static Control Systems/3M
- Wescorp

DIP TUBES
- Bengal
- Colvin

Dip Tubes *(continued)*
    Semtronics
    Static Control Systems/3M
    Thielex
    Wescorp

TOTE BOXES AND STORAGE BINS
Bemis
Bengal
Charleswater
Colvin
Conductive Containers
Controlled Static
Glen Mitchel
Herman Miller
Lewis Systems
LNP
NRD
Protecta Pack
Republic Packaging
Semtronics
Sentinel Foam
Simco
Stanley Vidmar
Static Control Systems/3M
Static Handling
Wescorp

AIR IONIZERS
Charleswater
Colvin
Clean Room
Controlled Static
Frontier
Herbert
NRD
Scientific Enterprises
SIMCO
Static Control Systems/3M
Wescorp

ELECTROSTATIC DETECTORS AND VOLTMETERS
ACL
Anderson
Chapman
Charleswater

Electrostatic Detectors and Voltmeters *(continued)*
    Clean Air
    Electro Tech
    Monroe
    NRD
    Scientific Enterprises
    Semtronics
    SIMCO
    Static, Inc.
    Static Control Systems/3M
    Trek
    Wescorp

CONDUCTIVE FLOOR SURFACES, MATS, AND FOOTWEAR
    Bemis
    Bengal
    Charleswater
    Controlled Static
    MISCO
    NRD
    Plastic Systems
    Scientific Enterprises
    Semtronics
    SIMCO
    Static, Inc.
    Static Control Systems/3M
    Techni Bag
    Wescorp

GARMENTS/CLOTHING
    Best Way Services (Araclean)
    Charleswater
    Clean Room
    Colvin
    Controlled Static
    SIMCO
    Wescorp

TOPICAL ANTISTATS
    ACL
    Controlled Static
    Merix
    Richmond Division of Dixico
    SIMCO
    Sprayway

Topical Antistats *(continued)*
  Tech Spray
  Wescorp

CONSULTANTS
  Stephen Halperin V. Associates
  Julie Associates, Inc.
  SAR Associates
  Don White Consultants, Inc.
  Reliability Sciences, Incorporated

ADDRESSES OF MANUFACTURERS AND SUPPLIERS

ACL, Inc.
1960 E. Devon Avenue
Elk Grove Village, IL 60007

Anderson Effects, Inc.
P.O. Box 657
Mentone, CA 92359

Armand Manufacturing, Inc.
725 Mateo Street
Los Angeles, CA 90021

Bemis Company
ESD Protective Materials Department
800 Northstar Center
Minneapolis, MN 55402

Bengal, Inc.
15226 Parthenia
Sepulveda, CA 91343

BestWay/Araclean
One North Beacon Street
La Grange, IL 60525

Chapman Corporation
P.O. Box 427
Portland, ME 04112

Charleswater Products, Inc.
93 Border Street
West Newton, MA 02165

Clean Room Products, Inc.
56 Penataquit Avenue
Bay Shore, NY 11706

Colvin Packaging Products, Inc.
1391 Hundley Street
Anaheim, CA 92806

Conductive Containers, Inc.
106 Wilmot Road
Suite 209
Deerfield, IL 60015

Controlled Static
9836 Jersey Avenue
Santa Fe Springs, CA 90670

Electro-Tech Systems, Inc.
35 E. Glenside Avenue
Glenside, PA 19038

Frontier Electronics, Inc.
Poinsett Highway
P.O. Box 625
Greenville, SC 29602

Gary Plastic Packaging Corporation
770 Garrison Avenue
Bronx, NY 10474

Glen-Mitch Tool, Inc.
722 W. Morse Street
Schaumburg, IL 60193

Stephen Halperin V. Associates
Professional Static Control Services
P.O. Box 1225
Elmhurst, IL 60126

Herbert Products, Inc.
180 Linden Avenue
Westbury, NY 11590

Herbert Miller, Inc.
Zeeland, MI

Jiffy Packaging
560 Central Avenue
Murray Hill, NJ 07874

Julie Associates, Inc.
P.O. Box 141
Billerica, MA 01821

Lewisystems, Menasha Corporation
426 Montgomery Street
Watertown, WI 53094

LNP Corporation
412 King Street
Malvern, PA 19355

Meridan Molded Plastics
P.O. Box 923
Meridan, CT 06450

Merix Chemical Company
2234 E. 75th Street
Chicago, IL 60649

Misco, Inc.
404 Timber Lane
Marlboro, NJ 07746

Monroe Electronics, Inc.
100 Housel Avenue
Lyndonville, NY 14098

NRD
2937 Alt Boulevard, N.
Grand Island, NY 14072

Nu-Concept Computer Systems, Inc.
Rt. 309 and Advance Lane
Colmar, PA 18915

Protecta-Pack Systems
705 Pennsylvania Avenue S.
Minneapolis, MN 55426

Reliability Sciences, Incorporated
2361 S. Jefferson Davis Highway
Suite M-111
Arlington, Virginia 22202

Republic Packaging Corporation
9160 S. Green Street
Chicago, IL 60620

Richmond Division of Dixico, Inc.
P.O. Box 1129
Redlands, CA 92373

SAR Associates
RR 2, Box 500
Rome, NY 13440

Scientific Enterprises, Inc.
2801 Industrial Lane, Box 220
Broomfield, CO 80020

Sealed Air Corporation
Old Sherman Turnpike
Danbury, CT 06810

Semtronics
P.O. Box 592
Martinsville, NJ 08836

Sentinel Foam Products
North Street
Hyannis, MA 02601

SIMCO Co., Inc.
2257 N. Penn Road
Hatfield, PA 19440

Stanley Vidmar
P.O. Box 1151
11 Grammes Road
Allentown, PA 18105

Static Control Systems/3M
223-2SW, 3M Center
St. Paul, MN 55101

Static Handling, Inc.
P.O. Box 671
Athens, OH 45701

Static, Inc.
P.O. Box 80
Skippack, PA 19474

Tech Spray
P.O. Box 949
Amarillo, TX 79105

Thielex Plastics Corporation
201 Eleventh Street
P.O. Box 518
Piscataway, NJ 08854

Trek
1674 Quaker Road
Barker, NY 14012

Vinyl Plastics, Inc.
P.O. Box 451
Sheboygan, WI 53081

Voyager Technologies, Inc.
6312 Variel Ave., Suite 211
Woodland Hills, CA 91367

Wescorp/DAL Industries, Inc.
1155 Terra Bella Avenue
Mountain View, CA 94043

Don White Consultants, Inc.
State Route 625, P.O. Box D
Gainesville, VA 22065

EOS/ESD Association, Inc.
P.O. Box 298
Westmoreland, NY 13490

Reliability Analysis Center
RADC/RAC
Griffiss Air Force Base, NY 13441

ESD PUBLICATIONS

U.S. GOVERNMENT DOCUMENTS

FED-STD-1018: Preservation, Packaging and Packaging Materials Test Procedures; Test Method 4046: Electrostatic Properties of Materials

PPP-C-1842: Cushioning Material, Plastic, Open Cell (For Packaging Applications)
Specification Sales (3FRSBS)
Bldg. 107, Washington Navy Yard
General Services Administration
Washington, DC 20407

DOD-HDBK-263: Electrostatic Discharge Control Handbook for Electrical and Electronic Parts, Assemblies and Equipment

DOD-STD-1686: Electrostatic Discharge Control Program for Electrical and Electronic Parts, Assemblies and Equipment

MIL-STD-129H: Marking for Shipment and Storage

MIL-STD-7588: Packing Procedures for Submarine Repair Parts

MIL-STD-8838: Test Methods and Procedures for Microelectronics: Test Method 3015: Electrostatic Discharge Sensitivity

MS-90363G: Box, Fiberboard, with Cushioning for Special, Minimum Cube Storage and Limited Reuse Applications

MIL-B117: Bags, Sleeves and Tubing, Interior Packing

MIL-S-19491: Semiconductor Devices, Packaging of

MIL-M-38510: Microcircuits, General Specification for

MIL-M-55565A: Microcircuits, Packaging of

MIL-B-B1705B: Barrier Materials, Flexible, Electrostatic-Free, Heat Sealable

MIL-P-81997A: Porches, Cushioned, Flexible, Electrostatic-Free Transparent

MIL-W-80C: Transparent Antielectrostatic Acrylic Base Observation Window

NAVSEA SE 003-AA-TRN-010: Electrostatic Discharge Training Manual
Commanding Officer
Naval Publications and Form Center
5801 Tabor Avenue
Philadelphia, PA 19120

INDUSTRY DOCUMENTS

AATCC Test Method 134-1975: Electrostatic Propensity of Carpets
American Society for Textile Chemists and Colorists
AATCC Technical Center, P.O. Box 12215
Research Triangle Park, NC 27709

ANSI Test Method 47 (Secretariat) 707 (Proposed): Electronic Devices
Sensitive to Electrostatic Discharges

ANSI Standard 241.3-1976: Conductive Safety-Toe Footwear
American National Standards Institute
1430 Broadway
New York, NY 10018

ASTM Test Method D257-58: D-C Resistance or Conductance of Insulating Materials

ASTM Test Method D991-75: Rubber Property-Volume Resistivity of Electrically Conductive and Antistatic Products

ASTM Test Method D2679-78: Electrostatic Charge

ASTM Test Method D3509-76: Electrostatic Strength Due to Surface Charges
American Society of Testing and Materials
1916 Race Street
Philadelphia, PA 19103

EIA Standard RS-471: Attention Symbol and Label for Electrostatic Sensitive Devices
Electronic Industries Association
2001 Eye Street
N.W. Washington, DC 20006

NAS 853: Field Force, Protection For
National Standards Association
1321 Fourteenth Street N.W.
Washington, DC 20005

NFPA Standard 56A: Inhalation Anesthetics; Section 46: Reduction in Electrostatic Hazard
National Fire Protection Association
470 Atlantic Avenue
Boston, MA 02110

Note: The above list was taken from EOS/ESD Association Newsletter, September 1983

Reliability Analysis Center
  1. Electrical Overstress/Electrostatic Discharge Symposium Proceedings
  2. ESD Protective Material and Equipment: A Critical Review, Spring 1982, SOAR-1

Electronic Industries Association

EIA Interim Standard No. T 5, "Packaging Material Standards for Protection of Electrostatic Discharge Sensitive Devices"

MAGAZINES

Evaluation Engineering
1281 Old Skokie Road
Highland Park, Illinois 60035

Microcontamination
2422 Wilshire Blvd
Santa Monica, CA 90403

EMC Technology
State Route 625
P.O. Box D
Gainesville, VA 22065

References

Chapter 1

1. Frank, Donald, "ESD Phenomenon and Effect on Electronics Parts, "Avionics Maintenance Conference ESD Seminar,"
2. Frank, D. E., "The Perfect '10'—Can You Really Have One?" EOS-3, 1981 Los Angeles, 1 April 1981
3. McMahon, E. J., T. N. Bhar, and T. Oishi, "Proposed MIL-STD and MIL-HDBK for an Electrostatic Discharge Control Program—Background and Status," EOS-1, 1979
4. Unger, Burt, ESD Tutorial, Orlando, Florida, 1982
5. Static Control Systems/3M News Release, 1980
6. Moss, Dick, Hewlett Packard, quoted at ESD Tutorial, Orlando, Florida, 1982
7. DeChairo, Lou, Bell Laboratories, quoted at ESD Tutorial, Orlando, Florida, 1982
8. Halperin. Steve, quoted in "Protecting ICs From Electrostatic Discharge," Electronic Products Magazine, June 1980, by Warren Yates
9. McFarland, W. Y., "The Economic Benefits of an Effective ESD Awareness and Control Program—An Empirical Analysis," EOS-3, 1981
10. Berbeco, G., "Passive Static Protection: Theory and Practice," EOS-2, 1980
11. Military Handbook DOD-HDBK 263, "Electrostatic Discharge Control Handbook for Protection of Electrical and

Electronic Parts, Assemblies and Equipment (Excluding Electrically Initiated Explosive Devices)" 2 May 1980

12. Chace, Susan, "Advice to Defenders of the Earth: Touch Large Animal Before Firing," Wall Street Journal, January 30, 1981

13. Halperin, Stephen, "Coping With Static Electricity— Volume 2," Evaluation Engineering

14. Lerro, Joseph P., Jr., "Static Electricity; A Problem in American Industry (part 2)," Design News, November 2, 1981

15. Drewes, Dave, Sencore Inc.

16. Unger, Burton A., "How not to 'zap' a circuit," Bell Laboratories Record, May/June 1982

17. "Disasters of another kind," Telephony, January 5, 1981

18. Lerro, Joseph P., Jr., "Static Electricity: A Problem in American Industry (Part 1), "Design News, Oct. 19, 1981

19. Strand, C. J., A. Tweet, and M. E. Weight, "An Effective Electrostatic Discharge Protection Program," EOS-4, 1982

20. Pierce, D. G., and D. L. Durgin, "An Overview of EOS Effects on Semiconductor Devices," EOS-3, 1981

21. Static Control Systems/3M, "Suggested Telephone Industry Practices for Static-Safe Handling of Static-Sensitive Printed Wiring Cards and Microelectronic Devices,"

22. Keller, J. K., "Protection of MOS Integrated Circuits from Destruction by Electrostatic Discharge," EOS-2, 1980

23. Anderson, Dan, "Antistatics: The Spark-Free Approach to Electrostatic Damage Prevention," EOS-2, 1980

24. Greason, W. D., G. S. P. Castle, and D. R. Hibbert, "Analysis of ESD Damage in JFET Preamplifiers," EOS-2, 1980

25. Domingos, H., "Basic Considerations in Electro-Thermal Overstress in Electronic Components," EOS-2, 1980

26. Domingos, H. and R. Raghavan, "Circuit Design for EOS/ESD Protection," EOS-4, 1982

27. Yates, Warren, "Protecting ICs from Electrostatic Discharge," Electronic Products Magazine, June 1980

28. Branberg, G., "Electro-static Discharge and CMOS Logic," EOS-1, 1979

29. Youn, S. Y., N. Hartdegen, and M. Sharp, "ESD Minimization Technique for MOS Manufacturing Final Test Area," EOS-4, 1982

30. Hulett, T. V., "On Chip Protection of High Density NMOS Devices," EOS-3, 1981

31. Clark, O. M., "Electrostatic Discharge Protection Using Silicon Transient Suppressors," EOS-1, 1979

Chapter 2

1. Bossard, P. R., R. G. Chemelli, and B. A. Unger, "ESD Damage from Triboelectrically Charged IC Pins," EOS-2, 1980

2. Yenni, Donald M., Jr., and James R. Huntsman, "The Deficiencies in Military Specification MIL-B-81705: Considerations and a Simple Model for Static Protection" EOS-1, 1979

3. Military Handbook DOD-HDBK 263, "Electrostatic Discharge Control Handbook for Protection of Electrical and Electronic Parts, Assemblies and Equipment (Excluding Electrically Initiated Explosive Devices), 2 May 1980

4. Private communication from Burt Unger, October 1983

5. Head, G. O., "Drastic Losses of Conductivity in Antistatic Plastics," EOS-4, 1982

6. Controlled Static brochure "Lightning Strikes"

7. Davenport, Donald E., "Metalloplastics," EOS-4, 1982

8. Krauskopf, Konrad B. and Arthur Beiser, "Fundamentals of Physical Science," McGraw-Hill, New York, 1971

9. Unger, B., R. Chemelli, P. Bossard, and M. Hudock, "Evaluation of Integrated Circuit Shipping Tubes," EOS-3, 1981

10. Youn, S. Y., N. Hartdegen, and M. Sharp, "ESD Minimization Technique for MOS Manufacturing Final Test Area," EOS-4, 1982

11. Kimball, Arthur L., "College Textbook of Physics," 1954, Sixth Edition, Revised by Alan T. Waterman, Henry Holt and Company

12. Halperin, S. A., "Facility Evaluation: Isolating Environmental ESD Problems," EOS-2, 1980

13. Keller, J. K., "Protection of MOS Integrated Circuits from Destruction by Electrostatic Discharge," EOS-2, 1980

14. Head, G. O., "A Low-Cost Program for Evaluation of ESD Protective Materials and Equipment," EOS-3, 1981

15. Antinone, R. J., "Microcircuit Electrical Overstress Tolerance Testing and Qualification," EOS-2, 1980

16. Berbeco, George R., "Static Protection in Electronic Manufacturing," Electronic Packaging and Production, July 1980

17. Berbeco, G., "Passive Static Protection: Theory and Practice," EOS-2, 1980

Chapter 3

1. Bolasny, R. E., "Static Control Systems," EOS-2, 1980
2. 3M Product Literature
3. Monroe Electronics Product Literature

4. DOD-HDBK-263, "Electrostatic Discharge Control Handbook for Protection of Electrical and Electronic Parts, Assemblies and Equipment (Excluding Electrically Initiated Explosive Devices", 2 May 1980

5. Monroe Electronics Product Literature and Instruction Manuals

6. Electro-Tech Systems, Inc. Product Literature

7. Private communication from Burt Unger

Chapter 4

1. Kolyer, John M. and William E. Anderson, "Selection of Packaging Materials for Electrostatic Discharge-Sensitive (ESDS) Items," EOS-3, 1981

2. Kanarek, Jess J., "As I See It . . .", Wescorp, Mountain View, Calif., Release to Circuits Manufacturing

3. Strand, C. J., A. Tweet, and M. E. Weight, "An Effective Electro-static Discharge Protection Program," EOS-4, 1982

4. Head, G. O., "Drastic Losses of Conductivity in Antistatic Plastics," EOS-4, 1982

5. Yenni, D. M., Jr. and J. R. Huntsman, "The Deficiencies in Military Specification MIL-B-81705: Considerations and a Simple Model for Static Protection," EOS-1, 1979

6. Lerro, Joseph P., Jr. "Static electricity: a problem in American industry (Part 1)," Design News, October 19, 1981

7. DOD-HDBK 263, 2 May 1980, "Electrostatic Discharge Control Handbook for Protection of Electrical and Electronic Parts, Assemblies and Equipment (Excluding Electrically Initiated Explosive Devices)

8. Giuliano, Jerry R., "Practical Views on Static Control," Lake Publishing

9. Phillips, L. P., "Basic Specification for ESD Protection in Industry," EOS-4, 1982

10. Texas Instruments, "Guidelines for Handling Electrostatic Discharge Sensitive (ESDS) Devices and Assemblies," May 24, 1982

11. Fuqua, Norman B., "ESD Protective Material and Equipment: A Critical Review," Spring 1982, SOAR-1

12. Anderson, Dan C., "Booming electronics industry spurs antistatic packaging," Package Engineering, February 1982

13. Unger, B., R. Chemelli, P. Bossard, and M. Hudock, "Evaluation of Integrated circuit Shipping Tubes," EOS-3, 1981

14. Huntsman, J. R. and D. M. Yenni, Jr. "Test Methods for Static Control Products," EOS-4, 1982

Chapter 5

1. Storm, D. C. "Controlling Electrostatic Problems in the Fabrication and Handling of Spacecraft Hardware," EOS-1, 1979
2. DOD-HDBK 263, 2 May 1980, "Electrostatic Discharge Control Handbook for Protection of Electrical and Electronic Parts, Assemblies and Equipment (Excluding Electrically Initiated Explosive Devices)
3. Fuqua, Norman B., "ESD Protective Material and Equipment: A Critical Review, "SOAR-1, Reliability Analysis Center, Spring 1982
4. Halperin, S. A., "Facility Evaluation: Isolating Environmental ESD Problems," EOS-2, 1980
5. Swenson, David E. and Donald M. Yenni, Jr., "Ionized Air for the Static Safe Environment," Presented at 1981 Nepcon/West Conference, Anaheim, Calif.
6. Keers, Joseph J. and Robert J. Kunz, "Nuclear Static Eliminators, Their Development and Uses," 3M Technical Information J-SNSE (1281) R1, no date
7. Herbert Curastat Production Information
8. Navy Environmental Health Bulletin, NAVMED P-5112
9. Bolasny, R. E., "Static Control Systems," EOS-2, 1980
10. Chapman Corp. Product Information
11. Halperin, S. A., "Static Control Using Topical Antistats," EOS-1, 1979

Chapter 6

1. Huntsman, J. R. and D. M. Yenni, Jr., "Test Methods for Static Control Products," EOS-4, 1982
2. Sohl, John E., "An Evaluation of Wrist Strap Parameters," EOS-2, 1980
3. Private communication from Hughes Aircraft to Norman B. Fuqua, referenced in "ESD Protective Material and Equipment: A Critical Review," Spring 1982, SOAR-1
4. DOD-HDBK 263, 2 May 1980, "Electrostatic Discharge Control Handbook for Protection of Electrical and Electronic Parts, Assemblies and Equipment (Excluding Electrically Initiated Explosive Devices)"
5. Texas Instruments, "Guidelines for Handling Electrostatic Discharge Sensitive (ESDS) Devices and Assemblies," May 24, 1982
6. Shelton, Scott E., "Why Use Antistatic Garments?" Evaluation Engineering, March/April, 1979
7. Burnett, Eric S., "ESD & Contamination From Clean Room Garments—Problems and Solutions," EOS-4, 1982
8. Halperin, Stephen A., "Facility Evaluation: Isolating Environmental ESD Problems" EOS-2, 1980

Chapter 7

1. Sheehan, D. K. and J. E. Burroughs, "Electrostatic Discharge Immunity in Computer Systems," EOS-4, 1982
2. DOD-HDBK-263, 2 May 1980, "Electrostatic Discharge Control Handbook for Protection of Electrical and Electronic Parts, Assemblies and Equipment (Excluding Electrically Initiated Explosive Devices)
3. Phillips, L. P., "Basic Specification for ESD Protection in Industry," EOS-4, 1982
4. Schnetker, Ted R., "Human Factors in Electrostatic Discharge Protection," EOS-1, 1979
5. Strand, C. J., A. Tweet, and M. E. Weight, "An Effective Electrostatic Discharge Protection Program," EOS-4, 1982
6. Kirk, Whitson J., Jr., "Control of Electrostatic Damage to Electronic Circuits," Bendix Report BDX-613-2428, March 1980
7. McAteer, Owen J., "An Effective ESD Awareness Training Program," EOS-1, 1979
8. Frank, Donald E., "The Perfect '10'—Can You Really Have One?" EOS-3, 1981
9. Molnar, Dave, as quoted in "Protecting ICs From Electrostatic Discharge," Electronic Products Magazine, June 1980, by Warren Yates

Glossary

1. Halperin, S. A., "Static Control Using Topical Antistats," EOS-1, 1979
2. McAteer, O. J., R. E. Twist, and R. C. Walker, "Latent ESD Failures," EOS-4, 1982
3. Unger, B., R. Chemelli, P. Bossard, and M. Hudock, "Evaluation of Integrated Circuit Shipping Tubes," EOS-3, 1981
4. Chase, E. W., "Evaluation of Electrostatic Discharge Damage to 16K EPROMS," EOS-3, 1981
5. Davenport, D. E., "Metalloplastics," EOS-4, 1982
6. Halperin, S. A., "Facility Evaluation: Isolating Environmental ESD Problems," EOS-2, 1980

Glossary

Like any other engineering discipline, electrostatic discharge has its own peculiar vocabulary. Because this field is so new, many of the terms used in it are defined here. While not all of these terms are used in this book, they are included here to allow the reader to use other materials on electrostatic discharge.

These definitions have been compiled from Electronic Industries Association (EIA) Interim Standard No. 5, January 1983; Military Handbook DOD-HDBK-263, 2 May 1980, Electrostatic Discharge Control Handbook for Protection of Electrical and Electronic Parts, Assemblies and Equipment (Excluding Electrically Initiated Explosive Devices); Electrical Overstress (EOS)/Electrostatic Discharge Symposium Proceedings; and industry brochures.

The definitions labeled EIA and DOD HDBK 263 have been edited slightly but are essentially the same as they appear in their parent documents; for the exact definition the reader should refer to those documents.

acetone—a highly flammable solvent commonly used in electronic parts/equipment manufacture. Wrist straps need to be resistant to such solvents.

AES—abbreviation for Auger electron spectroscopy.

air ionizer—a device for supplying positive and negative air ions to nonconductive items in order to neutralize electrostatic charges on them. It is a supplement to wrist straps and protective work surfaces, but not a substitute for them.

alpha particle—A positively charged particle consisting of two protons and two neutrons emitted from the nucleus during a decay process. Because of its size, it has poor penetrating ability. It cannot penetrate skin, but it can ionize air molecules. (EIA) It can also affect circuit performance. Alpha particles from ceramics have caused memory losses.

ampere—Unit of electrical current. One ampere of current is 6.24×10^{18} electrons passing one point in one second. Abbreviated as amp. 1 amp = 1 coulomb/second (EIA).

antistatic materials—Those materials that resist triboelectric charging and produce minimal static charges when separated from themselves or other materials. Impregnated types: Those materials impregnated with migratory antistats. Surface treated types: Those materials which have been treated by spraying, dipping, printing, or wiping with a topical antistatic agent to render them surface lubricious. ESD protective materials having a surface resistivity greater than 10^9 but not greater than 10^{14} ohms per square. (DOD-HDBK 263) A material is sufficiently conductive to be antistatic if it has a volume resistivity of 10^{11} ohm-cm maximum. (MIL STD 883B)

antistats—Liquids that make a material static-controlled when applied to it.

arborite—A type of ESD protective benchtop and laminate.

ASTM—Abbreviation for American Society for Testing Materials.

Auger electron spectroscopy—A method for examining materials and surfaces in very fine detail.

avalanche degradation—See "thermal secondary breakdown."

bag—A performed container made of flexible material generally enclosed on all sides except one that forms an opening which may or may not be sealed after loading. It is normally constructed from one piece of material that has been folded over and sealed on two edges. (EIA)

bipolar device—A semiconductor device that uses both positive and negative carriers in the circuit.

blue poly—Variation on pink poly antistatic polyethylene.

board shunt—See "shunting bar."

body capacitance—The capacitance that exists between the human body and the earth. Generally between 100 and 300 pF.

body resistance—The resistance of the human body as typically measured between the hands. Its value depends on several variables such as perspiration. For simulation of electrostatic discharge in test circuits, its value is set at 1500 ohms.

bonding—Interconnecting conductive parts so as to maintain a common electrical potential.

bulk breakdown—A power-dependent ESD-related failure mechanism in microelectronic and semiconductor devices and piezoelectric crystals. It results from changes in junction parameters due to high local temperatures within the junction area. Such high temperatures result in metallization alloying or impurity diffusion resulting in drastic changes in junction parameters. The usual result is the formation of a resistance path across the junction. This effect is usually preceded by thermal secondary breakdown. (DOD HDBK 263)

bulk resistivity—See "volume resistivity."

capacitance—The ability of a component or material to store an electric charge. The capacitance of a charged conductor is the ratio of its charge to its voltage (i.e., $C = Q/V$). Capacitance is measured in terms of "farads." Since the farad is such a very large number, capacitance is usually expressed in millionths of a farad or "microfarads" or millionths of millionths of a farad or "picofarads."

carrier—1. A holder for electronic parts and devices that facilitates handling during processing, production, imprinting, or testing operations and protects such parts under transport. (EIA) 2. A basic component of topical antistats which acts as a vehicle to transport the antistatic mechanism. It can be either water, alcohol, or another solvent.[1]

catastrophic failure—A sudden and complete failure, typically an open circuit or a short circuit.[2]

CDM—Abbreviation for charged device model.

charge—The electrical energy stored in a capacitor or on an insulated object. Measured in coulombs or fractions thereof. The static charge on a body is measured by the number of separated electrons on the body (negative charge) or the number of separated electrons not on the body (positive charge). Since electrons cannot be destroyed, an electron removed from one body must go to another body, leaving behind a positive (+) void. Thus, there are always equal and opposite charges produced. (EIA)

charged device model—The model of a device that acquires a charge, which is then discharged to ground through a low impedance path. The discharge takes a few nanoseconds or even less. Peak current may be 10 amps or more.[3,4]

charging wand—A moveable hand-held device used to induce a charge on a surface.

clean room—A room or area with elaborate equipment and procedures for suppressing dust particles in the air. Such rooms are required for assembly of microcircuits as dust particles can easily contaminate such circuits.

CMOS—Abbreviation for complementary MOS.

conductive materials—1. Materials having surface resistivities of 10^5 ohms per square or less such as metals, bulk conductive plastics, wire impregnated materials, and conductive laminates. (DOD HDBK 263). 2. Those materials which are either metal or impregnated with metal, carbon particles or other conductive materials, or whose surface has been treated with such through a process of lacquering, plating, metallizing, or printing. These materials seldom charge when separated from one another but can produce charging when separated from nonconducting surfaces. (EIA)

conductivity—A prime characteristic for providing protection against stationary or approaching charged bodies or people by limiting accumulation of residual voltages. (DOD HDBK 263) The ability to conduct charges.

conductor—A substance or body that allows a current of electrons to pass continuously along it or through it when a voltage is applied across any two points. Such materials exhibit relatively low resistance (high conductivity). (EIA)

corona—A bluish discharge from a conductor, particularly with sharply curved surfaces, that occurs when the voltage is high enough to ionize the surrounding air.

coulomb—A specific quantity of electrons (charge) on a body, expressed as 1 coulomb = 6.24×10^{18} electrons. In electrostatics a much more practical unit is the nanocoulomb 10^{-9} coulombs) representing a charge of 6.24×10^9 electrons. (EIA)

coulombmeter—An instrument for measuring the quantity of charge stored in a circuit or object.

Coulomb's law—One of the fundamental laws in electrostatics. The force of attraction or repulsion between two charges of electricity is proportional to the product of their magnitudes and inversely proportional to the square of the distance between them. If the charges are like (either both positive or both negative) the force will be repulsive. If the charges are unlike, the force will be attractive.

current—The flow of electrons past a certain point in a specified period of time, measured in amperes or fractions thereof. Current is measured in terms of electrons per second, but since this number would be tremendously large, it is usually stated in terms of "coulombs per second." 1 coulomb per second = 1 ampere. (EIA)

current limiting resistor—A resistor inserted in an electric circuit, often as a protective device, to limit the flow of current to some predetermined value.

decay time—The time for a static charge to be reduced to a given percent of the charge's peak voltage. It is an indirect method of measuring material surface resistivity as it is generally directly proportional to surface resistivity. (DOD HDBK 263)

degradation failure—A gradual and partial failure of a device that occurs when one of its parameters shifts outside its specified limits.

dielectric—A nonconducting material, an insulator, such as air, mica, and ceramic used to separate the plates of a capacitor.

dielectric breakdown—A voltage-dependent failure mechanism in microelectronic and semiconductor devices. When a potential difference is applied across a dielectric region in excess of the region's inherent breakdown characteristics, a puncture of the dielectric occurs. It can result in either total or limited degradation of the part depending on pulse energy. The part may heal. (DOD HDBK 263) A threshold effect in a dielectric medium where, at some electric field strength across the medium, bound electrons become unbound and travel through the medium as a current. In solid media, the region of the current is permanently damaged. The units of measurement are usually volts per unit of thickness. (EIA)

DIP—Abbreviation for dual-in-line package (semiconductors).

DIP sticks—See "magazine."

dual-in-line package—A type of housing for integrated circuits. The standard form of which is molded plastic with two rows of pins. These packages are also made of pre-molded ceramic with metal or ceramic ends or lids. It is a more popular type of containment than the flat pack or TO-can for industrial use because it is relatively inexpensive and easily wave soldered into printed circuit boards. (EIA)

EAP—Abbreviation for electroactive polymer.

edge protector—See "shunting bar."

electrical and electronic part—A part such as a microcircuit, discrete semiconductor, resistor, capacitor, thick or thin film device, or piezoelectric crystal. (DOD HDBK 263)

electrically continuous—A surface that is electrically conductive in that current can be passed as the result of an applied voltage between any two points on its physical surface and when discontinuities, slots, or holes do not occupy more than 10% of the material's surface. (EIA)

electrically powered static eliminator—Consists of two basic components: the static eliminator, which generally consists of one or more electrified needles rigidly held from a grounded metal housing or proximity ground, and the high-voltage supply which powers the static eliminator. Ion generation occurs in the air space surrounding the highly charged needle points. (EIA)

electroactive polymer—A conductive packaging material still in experimental stage. Abbreviated EAP.

electromagnetic compatibility—The ability of electronic equipment to operate as intended without creating unacceptable electromagnetic interference or respond to such interference beyond specified limits.

electromagnetic interference—An electromagnetic disturbance caused by static sparks, lightning, radar, radio and TV transmission, brush motors, line transients, etc. By line conduction or air propagation, EMI can induce undesirable voltage signals in electronic equipment causing malfunction and occasionally component damage. Protection against EMI usually requires the use of shields, filters, and special circuit design. See also "electromagnetic shield, electrostatic shield, and radio frequency interference." (EIA)

electromagnetic shield—A screen or other housing placed around devices or circuits to reduce the effect of both electric and magnetic fields on or by them. The electromagnetic field results from the presence of a rapidly moving electric field (rf) and its associated magnetic field. Shielding from electromagnetic interference (abbreviated EMI) is a combination of reflection and absorption of electromagnetic energy by the material. Reflection occurs at the surface much like the reflection of light at an air-to-water interface, and is not usually affected by shield thickness. Absorption, however, occurs within the shield and is highly dependent on thickness. The best electromagnetic shielding materials are ferritic, i.e., steel, nickel, etc. Aluminum is often used in less critical situations, however. (EIA)

electrometer—An instrument for measuring a potential difference or an electric charge. A typical use is static decay measurements.

electromigration—A failure mechanism in fine line structures caused by the mass transport of metal by momentum exchange between thermally activated metal ions and conducting electrons. The result is a thinning of metal in some regions (voids) and buildup in other regions (hillocks).

electron—A negatively charged elementary particle with an electrical charge equal to about 1.6×10^{-19} coulomb. (EIA)

electroscope—A primitive instrument used to detect an electric charge. When calibrated, it can measure potential difference. In its common form, it consists of two thin narrow strips of metal, such as gold leaf, that are suspended from a common point in a glass or metal container. When charged, the strips will spread apart. The separation between the strips is proportional to the charge.

electrostatic charge—An electric charge on the surface of an insulated object.

electrostatic detector—An instrument used to determine the presence or absence, polarity, and relative magnitude of electrostatic charges in the work area. If accurate measurements of such charges or static decay rates are needed, one must use the more elaborate, laboratory grade electrostatic voltmeter. Types of electrostatic detectors include electrometer amplifiers, electrostatic voltmeters, electrostatic field meters, and leaf detection electroscopes.

electrostatic discharge—A transfer of electrostatic charge between bodies at different electrostatic potentials caused by direct contact or induced by an electrostatic field. (DOD HDBK 263) A rapid transient.

electrostatic discharge sensitivity—The relative tendency of a device's performance to be degraded by ESD. (MIL-STD 883B). Abbreviated ESDS.

electrostatic discharge simulator—An instrument used to simulate the discharge of static electricity from the human body.

electrostatic field—1. The region surrounding an electrically charged object in which another electrical charge can be induced and will experience a force. Quantitatively, it is the voltage gradient between two points at different potentials. (EIA). 2. A voltage gradient between an electrostatically charged surface and another surface of a different electrostatic potential. (DOD-HDBK 263)

electrostatic fieldmeters—Instruments used to measure the electrostatic fields produced by charged bodies using a non-contact probe or sensor, and provide readings in electrostatic field strength or electrostatic voltage at a calibrated distance from a charged body. (DOD-HDBK 263)

electrostatic overstress simulator—See "electrostatic discharge simulator."

electrostatic sensitive—See "ESD sensitive items."

electrostatic shield—A barrier or enclosure that prevents the penetration of an electrostatic field. An electrostatic shield, however, may not offer much protection against the effects of electromagnetic interference (EMI). EMI shields, however, are good electrostatic shields. (EIA)

electrostatic shielding materials—Materials that are capable of attenuating or shunting an electrostatic field, so that its effects do not reach the stored or contained item and produce damage. (EIA) Typically a nickel or aluminum coated polyethylene packaging material that gives Faraday cage protection from external static fields and discharge.

electrostatics—The branch of physics that deals with electricity at rest.

EMC—Abbreviation for electromagnetic compatibility.

EMI—Abbreviation for electromagnetic interference.

energy—The ability to do or perform work. Measured in joules or fractions thereof. A spark is energy being expended. The measure of energy takes several forms. If it is electrical energy, it is measured in watt-seconds or joules. Since a joule is quite a bit of energy and since electric sparks do not usually have this much energy, their energy is usually measured in thousandsths of a joule (millijoule). A static spark in the order of millionths of a joule (microjoule) will damage semiconductors. Electrostatic energy is calculated from the relations $CV^2/2$, $Q^2/2C$ or $QV/2$. (EIA)

EOS—Abbreviation for electrical overstress.

equipotential bonding—Bonding that results in all conductors having the same potential.

ESD—Abbreviation for electrostatic discharge.

ESD protective material—Material capable of one or more of the following: limiting the generation of static electricity, rapidly dissipating electrostatic charges over its surface or volume, or providing shielding from ESD spark discharge or electrostatic fields. ESD protective materials are classified in accordance with their surface resistivity (or alternate conductivity) as conductive, static dissipative or antistatic. (DOD-HDBK 263)

ESD protective packaging—Packaging made with ESD protective materials to prevent ESD damage to ESDS items. (DOD-HDBK 263)

ESD sensitive items—1. Although all microelectronic devices are susceptible to ESD to some extent, for the purpose of this test method and the associated requirements, an ESDS device is one which can be damaged by exposure to ESD at a level less than 10,000 volts using the standard circuits shown in MIL-STD 883B. (MIL-STD 883B) 2. Electrical and electronic parts, assemblies and equipment that are sensitive to ESD voltages of 15,000 volts or less as determined by the test circuit given in DOD Handbook 263. (DOD-HDBK 263)

Faraday cage—An electrically continuous, conductive enclosure which provides electrostatic shielding. The cage is usually grounded although it need not be. (EIA)

finger cots—A latex or similar covering for fingers that prevents contamination of devices being handled in clean room conditions. Unfortunately, finger cots can generate static charges when rubbed against an object.

gaseous arc discharge—A voltage dependent ESD-related failure mechanism in microelectronic and semiconductor devices. For parts with closely spaced unpassivated-thin electrodes, gaseous arc discharge can cause degraded performance. The arc discharge condition causes vaporization and metal movement that is generally away from the space between the electrodes. (DOD HDBK 263)

gauntlet—A protective glove extending to the elbows. Used to cover long sleeved apparel which is not made of ESD protective material.

ground—1. A metallic connection with the earth to establish zero potential or voltage with respect to ground or earth. It is the voltage reference point in a circuit. There may or may not be an actual connection to earth, but it is understood that a point in the circuit said to be at ground potential could be connected to earth without disturbing the operation of the circuit in any way. Grounds that can be used for static control work stations include water pipes, any power ground, or any large metal structural member of a building. (EIA) 2. A mass such as earth, a ship or vehicle hull, capable of supplying or accepting a large electrical charge. (DOD HDBK 263)

ground, hard—A connection to ground either directly or through a low impedance. (DOD HDBK 263)

ground, soft—A connection to ground through an impedance sufficiently high to limit current flow to safe levels for personnel (normally 5 milliamperes). Impedance needed for a soft ground is dependent upon the voltage levels which could be contacted by personnel near the ground. (DOD-HDBK 263)

ground fault interrupter—A device that senses leakage current from faulty test equipment and interrupts the circuit almost instantaneously when these currents reach a potentially hazardous level. (DOD HDBK 263) Also called Ground Fault Circuit Interrupter.

grounded workbenches—Workbenches which contact ESDS items and personnel should have ESD protective work surfaces over the area where ESDS items would be placed. Workbench surfaces should be connected to ground through a ground cable. (DOD HDBK 263)

grounding—Connecting to ground or to a conductor that is grounded. A means of referencing all conductive objects to a zero voltage equipotential surface. This is the surest method of eliminating ESD since everything is maintained at the same potential. (EIA)

grounding strap—See "personnel ground strap."

handled or handling—Actions in which items are hand manipulated or machine processed during actions such as inspections, manufacturing, assembling, processing, testing, repairing, reworking, maintaining, installing, transporting, failure analysis, wrapping, packaging, marking or labeling. (DOD HDBK 263)

HBM—Abbreviation for human body model.

healing—In damaged semiconductors, the self-improvement of electrical characteristics.

HIC—Abbreviation for hybrid integrated circuit.

high voltage relay—A relay built to withstand high voltages on its contacts. It is used in many of the human body model static discharge simulators.

HMOS—Abbreviation for high density MOS.

human capacitance—See "body capacitance."

human resistance—See "body resistance."

humidity control—The process of adjusting the relative humidity in the work environment for minimizing electrostatic discharge. The higher the relative humidity (RH), the less will be the problems with ESD. However, high RH can make working conditions intolerable for the average person and create other manufacturing problems. Humidity control is not *the* solution to ESD problems; it must be used in conjunction with numerous other ESD control techniques.

humidity test chamber—An enclosure with controlled humidity used in performing static decay tests.

hygrometer—An instrument for measuring humidity of the atmosphere.

IFM—Abbreviation for ion flux monitor.

immobile charge—The charge residing on nonconductors.

induction—The process by which an electric charge establishes a charge in a nearby object without physical contact.

149

induction static eliminator—Instruments consisting of a series of conductive grounded points or brushes. When a single sharp grounded needle point is brought into the proximity of any highly-charged surface, it has induced in it a charge opposite that of the surface. When a high enough charge concentration has been developed, the surrounding air will break down. A vast number of charge balancing ions are formed. The simple "tinsel" static eliminator is an example of an induction static eliminator. (EIA)

input protection network—Resistors, capacitors, diodes, or transistors or any combination of these devices placed at the input of an integrated circuit to protect it from electrostatic discharge.

insulative material—A material having a volume resistivity of 10^{12} ohm-cm minimum, or a surface resistivity of 10^{14} ohms/square minimum. (MIL-STD 883B)

insulator—A material that does not conduct electricity. A nonconductor. (EIA)

integrated circuit—Any electronic device that contains transistors, resistors, capacitors, etc., in a single package. To further define the nature of an integrated circuit, additional modifiers are often prefixed. For instance, a bipolar IC is one containing junction semiconductors that use both positive (p-type) and negative (n-type) charge carriers. Abbreviated IC. (EIA)

ion flux monitor—An apparatus for determining the amount of air ions available from an ionizing source. Abbreviated IFM.

ionization—The process by which a neutral atom or molecule, such as air, acquires a positive or negative charge. (EIA)

isopropanol—A commonly used solvent in the electronics industry.

joule—A unit of energy. The energy of a static discharge is $\frac{1}{2}CV^2$ where C = capacitance of the discharging object and V = voltage difference between the discharging object and the point to which it discharged. (EIA)

latent failure—A failure that occurs in a device some time after it has been exposed to electrostatic discharge. At the time of the discharge there was no apparent damage.

lubricity—A measure of surface smoothness and lubricating action of moistness. Triboelectric generation is a friction process, the higher the lubricity of the surfaces being rubbed, the lower the friction and hence the lower the generated charges. (DOD HDBK 263)

magazine—A rigid package for shipping and storage of integrated circuits. Also called DIP stick, slide, or rail.

metallization melt—A power dependent ESD-related failure mechanism in microelectronic and semiconductor devices. It occurs when ESD transients increase part temperature sufficiently to melt metal or fuse bond wire. (DOD-HDBK 263)

metalloplastics—Plastics which have metallic fibers added. The composite has a low resistivity, thereby reducing the inherent problem of ESD in plastics.[5]

microlightning—Electrostatic discharge.

MIS—Abbreviation for metal-insulator-semiconductor.

MNOS—Abbreviation for metal nitride oxide semiconductor.

mobile charge—The charge resting on the conductive parts of an object.

MOS—Abbreviation for metal-oxide-semiconductor.

MOS device/structure—1. A device having a metal-oxide semiconductor layer structure. The oxide is a thin dielectric that is voltage sensitive and can be easily damaged by a static discharge or by induction from an electrostatic field. The damage occurs as a dielectric breakdown when the voltage or electric field across the dielectric layer between two relatively conductive layers exceeds the dielectric strength of that material. (EIA) 2. A conductor and a semiconductor substrate separated by a thin dielectric. Thus the acronym MOS for metal-oxide-semiconductor is derived. A more general acronym for this structure is MIS for metal-insulator-semiconductor. (DOD-HDBK 263)

MOSFET—Abbreviation for metal-oxide semiconductor field-effect transistor.

nanocoulomb—One billionth (10^{-9}) of a coulomb. As an example, maximum immobile charge measured on a DIP was 0.08 nanocoulomb.[3]

nickel 63—A weak beta emitter used in Trek's electrostatic voltmeter.

NMOS—Abbreviation for N-channel MOS.

nonconductor—See "insulator."

nuclear ionizer—See "nuclear static eliminator."

nuclear static eliminator—A device used to create ions by the irradiation of the air molecules. Some models use a safe alpha-emitting isotope to create sufficient ion pairs to neutralize a charged surface. The high speed particle interacts with air molecules with sufficient energy to actually strip off one of its outer electrons (see *ionization*). (EIA)

ohm—The unit of electrical resistance. It is the resistance through which a current of one ampere will flow when a voltage of one volt is applied. (EIA)

ohms per square—The unit of surface resistivity. For any material it is numerically equal to the surface resistance between two electrodes forming opposite sides of a square. The size of the square is immaterial. It is normally used as a resistivity measurement of a thin conductive layer or material over a relatively insulative base material. (DOD HDBK 263)

ozone—Triatomic oxygen. An unstable colorless or pale blue gas with a pungent characteristic odor. It can cause deterioration of rubber insulation on electrical conductors. At concentrations higher than 0.1 ppm it can be hazardous to people. It is produced as an undesired byproduct during ionization of air by high-voltage transformer ionizers. The amount produced by such ionizers is generally at a safe level.

package (device encapsulation)—In the electronics industry, packaging refers to the process of locating, connecting, and protecting various devices, components, etc. For example, entire circuits can be printed onto thin film wafers that are electrically interconnected during fabrication. These miniature circuits are then fused into an easy-to-handle element called a package. This package both protects the semiconductor circuitry and permits convenient external connections to be made to it. (EIA)

package (protection and distribution)—The enclosure of products, devices, or other packages in a wrap, pouch, bag, slide, magazine, or other container form to perform one or more of the following functions: (1) Containment for handling, transportation and use. (2) Preservation and protection of the contents for the life of the item. (3) Identification of contents including quantity and manufacturer. (4) Facilitate the dispensing and use of the contents. (EIA)

packing density—In an integrated circuit, the number of gates per unit area.

packaging materials—Those materials which cushion, enclose or protect the finished product during transportation and storage, such as bags, boxes, wraps, cushioning materials, foams, magazines (slides, tubes, rails). (EIA)

PCB—Abbreviation for printed circuit board.

personnel apparel—Recommended apparel for people handling ESDS items is long sleeved protective smocks or close-fitting, short sleeved shirts or blouses. (DOD HDBK 263)

personnel ground straps—Personnel handling ESDS items should wear a skin-contact wrist, leg or ankle ground strap. The function of such straps is to rapidly dissipate personnel static charges safely to ground and equalize personnel static levels with that of the work surface. (DOD HDBK 263) See also "wrist strap."

pink poly—Short for pink polyethylene which is a plastic impregnated with antistats.

plastic bubble pack—A type of protective packaging.

PMOS—Abbreviation for P-channel MOS.

potential—The degree of electrification. Measured in millivolts, volts, or kilovolts. Potential or voltage is measured from a base point. This point can be any voltage but is usually ground, which is theoretically zero voltage. (EIA)

pouch—A small or moderately-sized bag-like container constructed by the sealing on three edges of two flat sheets of flexible material or by sealing one end of a tube of flexible material. (EIA)

protected area—An area which is constructed and equipped with the necessary ESD protective materials and equipment to limit ESD voltage below the sensitivity level of ESDS items handled therein. (DOD HDBK 263)

protective handling—Handling of ESDS items in a manner to prevent danger from ESD. (DOD HDBK 263)

protective flooring—Available in the form of conductive, static dissipative and antistatic carpeting, vinyl sheeting, vinyl floor tiles and terrazzo. (DOD HDBK 263)

PVC—Abbreviation for polyvinyl chloride.

radioactive ionizers—See "nuclear static eliminator."

radio frequency interference—A form of electromagnetic interference (EMI). Any electrical signal capable of being propagated and interfering with the proper operation of electrical or electronic equipment. The frequency range usually includes the entire electromagnetic spectrum. The spark from a static discharge is a source of rf interference. Abbreviated RFI. (EIA)

resistance—The difficulty an electrical current encounters in passing through an electrical circuit or conductor. It is a bulk property of a material that depends on the material's dimensions, electrical resistivity, temperature, and also voltage in non-ohmic materials. The resistance of the material determines the current (electron flow) produced by a given voltage. The practical unit of resistance is the ohm. (EIA)

resistivity—A measure of the intrinsic ability of a material to conduct current. Its value is independent of the dimensions of the material. Both conductors and non-conductors have resistivity. The unit of volume resistivity is the ohm-cm. The unit of surface resistivity is ohms per square. (EIA)

SAW—Abbreviation for surface acoustic wave device.

SDP—Abbreviation for static discharge pulse.

SEM—Abbreviation for scanning electron microscope.

sensitive devices—See "ESD sensitive items."

sensitivity—The minimum value that a sensor will effectively and reliably detect. (EIA)

sheet resistivity—See "surface resistivity."

shielding—See "electrostatic shielding."

shipping tubes—See "magazine."

shuffle test—A test to evaluate the static generation properties of flooring materials. It involves sliding a foot across the floor.[6]

shunting bar—A device for shorting together the terminals of ESDS items using metal shunting bars, metal clips, or non-corrosive conductive foam. Also called "edge protector" or "board shunt." (DOD HDBK 263)

smock—A garment worn over street clothes to control charge on clothing.

SOS—Abbreviation for silicon-on-sapphire.

spark—An abrupt, short duration electric discharge that causes a flash of light. The electromagnetic pulse caused by ESD discharge in the form of a spark can cause part failure and cause equipment such as computers to upset. (DOD-HDBK 263)

Speidel band—A type of wrist strap named after the Speidel Co.'s wrist-watch strap. It provides dependable contact with the user's skin, making an effective body ground, which is important in ESD control.

static awareness—A desired state of mind for people who work with ESDS devices and components, whether in manufacture, assembly, test, or repair. Because they are convinced of the problem, they willingly follow special procedures and wear special clothing around ESDS devices.

static dissipative materials—ESD protective materials having surface resistivities greater than 10^5 but not greater than 10^9 ohms per square. Static dissipative materials could include the same materials as conductive materials except that the thicknesses are lower, wire or wire mesh included therein is finer or there is more space in conductive materials, or volume resistivities are higher. (DOD HDBK 263)

static electricity—electricity that is not moving.

static shielding—See "electrostatic shielding."

surface breakdown—A voltage-dependent ESD-related failure mechanism in microelectronic and semiconductor devices. For perpendicular junctions it is a localized avalanche multiplication process caused by narrowing of the junction space charge layer at the surface. (DOD HDBK 263)

surface resistivity—1. The ratio of dc voltage to the current that passes across the surface of the system. In this case, the surface consists of the geometric surface and the material immediately in contact with it. In effect, the surface resistivity is the resistance between the two opposite sides of a square and is independent of the size of the square or its dimensional units. Surface resistivity is relevant only for materials where current conduction is virtually only on the surface. Surface resistivity is not meaningful for volume conductive materials. The unit of surface resistivity is ohms per square. The unit ohms is occasionally used since the value is the resistance across a square. The use of ohms per square is recommended to distinguish surface resistivity from arbitrary resistance. (EIA). 2. An inverse measure of the conductivity of a material and equal to the ratio of the potential gradient to the current per unit width of the surface, where the potential gradient is measured in the direction of current flow in the material. For any material it is numerically equal to the surface resistance between two electrodes forming opposite sides of a square. The size of the square is immaterial. Surface resistivity applies to both surface and volume conductive materials and has the value of ohms per square. It normally is used as a resistivity measurement of a thin conductive layer or material over a relatively insulative base material. It is not constant for a homogeneous material but varies with material thickness. Therefore, the relationship of surface resistivity to volume resistivity is meaningless for a homogeneous bulk conductive material unless the thickness is also given. It is used to measure the resistivity of surface conductive materials such as hygroscopic antistatic polyethylenes, nylon, virgin cotton, metal or carbon coated paper plastics, and other conductively coated or laminated insulative materials. (DOD HDBK 263)

thermal secondary breakdown—A power-dependent ESD-related failure mechanism in microelectronic and semiconductor devices. Also known as avalanche degradation. Since thermal time constants of semiconductor materials are generally large compared with transient times associated with ESD pulses, there is little diffusion of heat from the areas of power dissipation and large temperature gradients can form in the parts. Localized junction temperatures can approach material melt temperatures, usually resulting in development of hot spots and subsequent junction shorts due to melting. This phenomenon is termed thermal secondary breakdown. (DOD HDBK 263)

transient suppressors—Devices that can reduce the voltage and energy flowing into an electrical circuit to levels sufficiently low to avoid damage to parts at the assembly levels. Suppressors include tin, zinc, or bismuth oxide voltage-dependent resistors (VDRs), often referred to as metal-oxide varistors, silicon voltage limiters, RC networks, and selenium stacks. (DOD HDBK 263)

triboelectric—Pertaining to an electrical charge generated by frictional rubbing or separation of two charges. (EIA)

triboelectric series—A list of substances in an order of positive to negative charging as a result of the triboelectric effect. (DOD HDBK 263) 2. A list of substances arranged so that any of them can become positively charged when rubbed with one farther down the list or negatively charged when rubbed with one farther up the list. Generally, the farther apart such materials are in the triboelectric series, the greater their tendency to charge one another. This series is derived from specially prepared and cleaned materials tested in very controlled conditions. In everyday circumstances, materials reasonably close to one another in the series can produce charge polarities opposite to that expected. This series is only a guide. (EIA)

trichloroethane—A common solvent used in the electronics industry. Wrist straps need to have some resistance to such solvents.

uniforms—In the world of ESD, commercially-cleaned company-provided static controlled garments for use in clean rooms.

upset failure—An intermittent failure caused by ESD. (DOD HDBK 263)

Van de Graaff generator—A very high voltage electrostatic generator with important uses beyond static electricity experiments.

VDR—Abbreviation for voltage-dependent resistor.

VLSI—Abbreviation for very large scale integration.

VMOS—Vertical groove MOS.

volt—The unit of voltage, potential, and electromotive force. One volt will send a current of one ampere through a resistance of one ohm. (Abbreviated V). (EIA)

voltage—The electrical potential difference that exists between two points and is capable of producing a flow of current when a closed circuit is connected between the two points. (EIA)

voltage-dependent resistor—A transient suppressor often referred to as metal-oxide varistor. Abbreviated VDR. (DOD-HDBK 263)

voltage suppression—A phenomenon where the voltage from a charged object is reduced by increasing the capacitance of the object rather than decreasing the charge on the object. The relation $Q = CV$ describes the phenomenon. It occurs most frequently when a charged object is close to a ground plane, but not in resistive contact with the ground plane. (EIA)

volume resistivity—1. The ratio of the dc voltage per unit of thickness, applied across two electrodes in contact with or imbedded in a specimen, to the amount of current per unit area passing through the system. Volume resistivity is generally given in ohm-centimeters. (EIA) 2. An inverse measure of the conductivity of a material and is equal to the ratio of the potential gradient to the current density, where the potential gradient is measured in the direction of current flow in the material. (Note: In the metric system, volume resistivity of an electrical insulating material in ohm-cm is numerically equal to the volume resistance in ohms between opposite faces of a 1 cm cube of the material. It is a constant for a given homogeneous material. Also referred to as bulk resistivity. (DOD HDBK 263)

wound—A degradation (in contrast to hard failure) of semiconductor components.

wrist strap—A conductive material circling the wrist in the fashion of a watch band. It is connected to ground through a current-limiting resistor. By conducting body charges to ground, it provides a first line of defense against ESD damage.

Index

Index

A

Air ionizers, 90-96
Amber, 21
Antistatic materials, 29, 73
Antistatic treatment techniques, 106
Antistats, 27
Atomic structure, 21-22
Atoms, 21

B

Bacteriostatic, 99
Bound charges, 47

C

Capacitance, 45-48
Capacitance, human body, 48
Capacitance effects, 48
Capacitor, 45
Capacitors, discrete, 46
CMOS, 4
Conduction, 32
Conductive materials, 29
Conductive shunts, 81
Conductors, 27-3223
Corona discharge, 44
Coulomb, 23
Coulomb, Charles, 23
Coulomb's, law, 23, 35

D

Dielectric breakdown, 12
Dielectric constant, 47

Dielectrics, 46-47
DIP sticks, 78DIP tubes, 74-78
Discharge, corona, 44
Discharge, electrostatic, 20

E

ECL, 2
Electrical overstress, 13
Electric field, 34-37
Electric potential, 41
Electromagnetic pulse, 9
Electronic components, effect of ESD on, 10-16
Electronic Industries Association (EIA), 31
Electrons, 21
Electron theory, 21-22
Electroscope, 38-41
Electroscope, gold-leaf, 38
Electrostatic attraction, 22-24
Electrostatic detectors, 53
Electrostatic discharge, 2-4, 20
Electrostatic fieldmeters, 55, 62-64
Electrostatic generators, 48-52
Electrostatic measurements, 38-41
Electrostatic meters, portable, 58-59
Electrostatic monitors, 53, 64-65
Electrostatic repulsion, 22-24
Electrostatic test equipment, 53-68
Electrostatic voltmeter block diagram, 60
Electrostatic voltmeters, 55, 59-62

Equipotential surface, 43
ESD, 2, 3, 6
ESD control program, 109-122
ESD damage, 8
ESD protective clothing, 104-108
ESD protective floor mat, 89
ESD protective floors, 87-89
ESD protective materials, 73, 98
ESD protective table mat, 88
ESD sensitive parts, classification of, 11
ESD sensitivity, 10
ESD simulator, human body, 50-52
ESDS, 2, 3
ESDS items, general guidelines for handling, 114-117

F
Failure, catastrophic, 15, 16
Failure, degradation, 16
Failure, delayed, 14
Failure, direct, 16
Failure, immediate, 14
Failure, indirect, 16
Failure, latent, 14
Failure, upset, 15
Faraday cage, 36
Force, lines of, 36

G
Gilbert, William, 21
Ground potential, 86-87
Grounded wrist straps, 100-104
Grounding, 85-86
Grounding considerations, 84-85
Grounds, hard, 32
Grounds, soft, 32

H
Handling procedures, 117-121
Human body ESD simulator, 50-52
Humidity, effects of, 83
Humidity control, 82-83
Humidity test chamber, 67-68
Hygroscopic agents, 30

I
Induction, 32-34
Inspection, 118
Inspection and test, 118-119
Insulators, 27-32
Integrated circuits, protection for, 16-19
Ionizing air guns, 96

J
JFET, 2

Junction burnout, 12

M
Material receiving area, 119
Metallization melt, 13
Meter, static decay, 67
Microscope, scanning electron, 13
MOSFET, 2, 20

N
Negative ion, 31, 90
Neutral body, 23
Neutrons, 21
NMOS, 1, 19

O
Occupational Safety and Health Act, 92
Ozone, 92

P
Packaging, protective, 69-81
Packaging area, 120-121
Personnel training, 121-122
PMOS, 1, 19
Positive ion, 31, 90
Potential difference, 42
Product design, 122
Protected areas, 111-114
Protection circuit, resistor-diode, 18
Protection networks, 17-19
Protective foam, 80-81
Protective packaging, 69-81
Protective tote boxes, 78-79
Protons, 21

R
Radioactive static eliminators, 95
Resistance, 28

S
SAW, 2
SCR, 2
Series resistor, 102
Shipping area, 120-121
Static charge dissipative, 51, 52
Static decay meter, 67
Static dissipative materials, 29, 83
Static electricity, 7-9
Storage bins, 78-79
Storeroom area, 119
Surface charge density, 44
Surface conductivity, 97
Surface lubricity, 97
Surface resistivity probe, 65-66

T
Test equipment, electrostatic, 53-68

Tools, 96-97
Topical antistats, 97-99
Transient suppressors, 16-17
Triboelectric charge, 27
Triboelectric charging, 24-27
Triboelectric effect, 24, 26
Triboelectric series, 26
TTL, 2

V
Van de Graaff renerator, 48-50
Very large scale integrated circuits, 1
Volume resistivity, 28

W
Work surfaces, 83-87